LYING DOWN

Other Works by the Same Author

As Fred Vassi

Why Is a Crooked Letter - plays
The Mysterious Familiar - poems

As Marco Vassi

The Stoned Apocalypse - memoir
The Metasex Manifesto (The Erotic Comedies) - fables and essays

Latent Image - novel

THE OLYMPIA QUARTET
Mind Blower - novel
The Gentle Degenerates - novel
The Saline Solution - novel
Contours of Darkness - novel

THE PORN QUARTET
The Devil's Sperm Is Cold - novel
Pro Ball Groupie - novel
In Touch - novel
Slave Parlour - novel

THE RELATIONSHIP DUET
The Sensual Mirror - novel
The Other Hand Clapping - novel

COLLECTION
Pieces for Bread - stories, essays, manifestos, journalism

Marco Vassi

LYING DOWN

The Horizontal Worldview

A Capra/Catalyst Book™ Capra Press, Santa Barbara, 1984

A Capra/Catalyst Book published by Capra Press

Vassi, Marco 19 -
Lying down: the horizontal worldview
1.
ISBN 0-88496-207-5
Library of Congress card number 84-1712

The author gratefully acknowledges permission to reprint excerpts from the following:

"Ching Chow." Reprinted by permission: Tribune Company Syndicate, Inc.
The Gravity Guiding System, *printed by permission from THE GRAVITY GUIDING SYSTEM® (originally Cum Gravity) by Robert Martin, M.D., published by Gravity Guidance, Inc. Copyright 1975 by Gravity Guidance, Inc.*
Three Worlds of Therapy *by Anthony Barton by permission of Mayfield Publishing Company. Copyright 1974 by Anthony Barton.*
Megatrends *by John Naisbitt by permission of Warner Books, New York.*
"Visions of Languor" by Linda Nochlin by permission of Linda Nochlin.
The Summing Up *by W. Somerset Maugham. Copyright 1938 by W. Somerset Maugham. Reprinted by permission of Doubleaay & Company, Inc.*
Sleep Positions *by Samuel Dunkell, M.D. Copyright 1977 by Delruth Incorporated. By permission of William Morrow & Company.*
Entropy *by Jeremy Rifkin. Copyright 1980 by Foundation on Economic Trends. Reprinted by permission of Viking Penguin, Inc.*
Phenomenological Psychology *by Erwin W. Straus. Copyright 1966 by Erwin W. Straus. Reprinted by permission of Basic Books, Inc., Publishers.*
"Physical" by Olivia Newton-John. Copyright 1981 by APRIL MUSIC, INC., STEPHEN A. KIPNER MUSIC, & TERRY SHADDICK MUSIC. Used by permission. All rights reserved.
The Network of Thought *by J. Krishnamurti. Reprinted by permission of Harper & Row.*
Holy the Firm *by Annie Dillard. Reprinted by permission of Harper & Row.*
Smut: Erotic Reality/Obscene Ideology *by Murray S. Davis, reprinted by permission of The University of Chicago Press.*
"The Heart is the Self" reprinted by permission of The Laughing Man Institute, P.O. Box 3680, Clearlake, CA 95422. Copyright The Laughing Man Institute. 1983.
Prisoner Without a Name, Cell Without a Number *by Jacobo Timerman, translated by Tony Talbot, reprinted by permission of Alfred A. Knopf, Inc.*
I Send A Voice *by Evelyn Eaton. Reprinted by permission of Quest Books, published by The Theosophical Publishing House.*
"Intimacy," by Taizan Maezumi Roshi, by permission of the Zen Center of Los Angeles.
The Dancing Wu Li Masters *by Gary Zukav, by permission of William Morrow & Co.*

Printed in USA

Designed by Kathleen Trainor Joynes: Neumeier Design Team
Illustrated by Dean Foster

To Shirley Winston and Alec Rubin,
my first and greatest teachers
of the craft of lying down.

To Tesora
for contribution beyond measure
in making this book possible.

To Fred Wertz
for sharing the birth of the vision
and for spectacular scholarship.

To Laura Archera Huxley
for Rainbow Walk.

To Mary Veronica Antonakos
for the connection.

Contents

Introduction 9

1
Position
The Physical Field 17

2
Exploration
The Intimate Field 38

3
Translation
The Metaphoric Field 53

4
Horizon
The Mystic/Absurd Field 92

Confession 98

CHING CHOW®
TRULY, 'TIS BETTER TO WALK THAN RUN, BETTER TO STAND THAN WALK, BETTER TO SIT THAN STAND, BETTER TO LIE THAN SIT.

Introduction

Perhaps the first thing to mention at the beginning of this book about lying down is that I am sitting at the typewriter as I compose these sentences. Most of the material that follows was gathered from the prone, using pad and pencil and tape recorder, but when it came to transcribing the tapes and notes, a typewriter was necessary. And I have found it impossible to type while lying down.

The fact that I am sitting while writing a work about lying down points to a major theme of this book, which has to do with the awakening of postural awareness, and awareness altogether. For example, what are you doing now? Are you standing, perhaps in the aisle of a book store? Or strolling on a beach? Or sitting in a comfortable chair? Or even lying on a soft couch as you read about lying down? I ask this question not as a self-remembering exercise but to suggest that your reading and understanding will be influenced not only by your posture, but by the body's awareness of itself in any posture. By extension, of course, this proposition covers not only reading but anything you do in life.

This idea is by no means new. It has been central to Eastern disciplines for thousands of years, and more recently been rediscovered by Western schools of healing, self-exploration and therapy. In some esoteric groups, it is even considered the very foundation of spiritual practice.

To take just one example: the postures referred to in the second paragraph—standing, sitting, walking and lying down—are called the "four dignities" in Buddhism. This is based on the teaching that to inhabit one's posture with real dignity, i.e., mindfulness and compassion, *is* enlightenment. Shunryu Suzuki made that quite plain in *Zen Mind, Beginner's Mind.*

However, he represented a very late stage in his tradition, a time in which sitting had become practically synonymous with postural awareness—standing and walking having been reduced to the status of fillers in the zendo, and lying down all but abolished as a form of weakness. So for Suzuki, sitting was "boss." And he was right, but only in the context of the worldview in which his tradition was embedded. Sitting is unquestionably a sublime and effective practice but, like Newtonian physics, its validity is limited to a certain scale of perception. Zazen, and all the sitting meditations, reflect a stage in human evolution when our species' verticality was beginning to be expressed in hierarchical and patriarchal systems.

That stage and those systems are coming to a turbulent end. The movement of the dharma to the West is itself part of a far larger transformation, a global upheaval that is making waves across the full spectrum of human culture, including religion, art, science, technology, philosophy, medicine, social relationships, and all the rest of it. We are witnessing a planetary spasm of species consciousness in which all our identities are being destroyed so that some new awareness might come into being.

Such a radical and universal change couldn't go unnoticed, and indeed one or another aspect of it has been described by various scholars and commentators. The term for it that I like best, and which I'll use throughout the book, comes from physics—the much-heralded "new paradigm." The word paradigm is derived from Greek and means "pattern, example, model." It came into prominence with the publication of Thomas Kuhn's *The Structure of Scientific Revolutions*, in which he notes that each scientific worldview is periodically overturned by a "paradigm shift." This doesn't refer merely to changes within an established system, but to the kind of fundamental redefinition of our understanding of the universe as was brought about by relativity and quantum mechanics.

Since the appearance of Kuhn's work, however, the concept of paradigm shift has been broadened. Ken Wilber, for example, states that for any paradigm to be complete it would have to unite science, philosophy-psychology, and religion-mysticism. And Marilyn Ferguson succinctly defines paradigm as "a collective map for the species."

Now, the relationship between postural awareness and the new paradigm might seem obscure, but it becomes clear when the two terms are placed in the proper equation, one which is somewhat startling on the face of it. This postulates that lying down *is* the new paradigm and, further, that lying down is the only radically new paradigm since our inception as a species. My argument here is that from the moment, several million years ago, when we altered our destiny by standing erect, until today, we have been shaped by the single paradigm of verticality, and the "shift" that is required now, quite simply, is going horizontal.

This may seem a whimsical idea, and it certainly has its amusing connotations, but the suggestion is perfectly serious, for lying down is a far more dimensional activity and concept than is ordinarily understood. In the same way that verticality has levels, horizontality has fields, which support and sustain the full range of thought, feeling and behavior in the human condition.

Following is a brief description of the four fields of lying down, each of which is explored more fully in successive chapters:

The first is the *physical*. This refers to the existential posture itself, the body reclining, with no necessary implications as to purpose or meaning. It is the sleeping body of Brahma whose dream the universe may be. It is the uncarved block.

The second is the *intimate*. This is the field of gnosis, or supple, intuitive knowing. Intimacy cannot be "practiced," but it can be uncovered and nourished through the posture of lying down, and the full breathing, relaxation, clarity and sensitivity that it engenders.

The third is the *metaphoric*. This relates to the attitude of horizontality as it applies to every category of human experience, insight and culture. The metaphor pervades everything from architectural styles to manners, from the pace of civilization to prayer.

The fourth is the *mystic/absurd*. This is accepting the truth of the horizon as absolute, acknowledging that we are bounded by impenetrable and unpassable limitations—the infinitely small, the infinitely large, the eternally enduring, the eternally changing, and the irreducible mystery of the existence of anything at all...the whole show curtained by the mortal horizons of birth and death. Having got that far, the mystic claims to transcend all such limitations, while the absurdist makes peace with the walls and finds solace in irony.

All of this relates to the human condition; but lying down, or assuming the horizontal, is the most basic attitude of life in general. The first life forms in the primeval soup assumed the horizontal, floating like webs on the water. And upon solidifying and growing different kinds of bodies, the subsequent fish retained the same orientation, as did the amphibians who ventured onto the land, and the lizards after them, and the vast majority of insects and mammals. The birds went quasi-erect, but when they take their verticality to the limit by flying, which is their essence, they resume the horizontal.

Our own "rise" as a species has literally to do with abandoning some three billion years of organic wisdom to try a most radical experiment—going vertical. The trees did it and were extremely successful, but that is because trees are stationary with a broad root base. However, the creatures who clambered about in their limbs attempted something more ambitious, to stay erect through sheer balance and also to exercise total freedom of movement.

It was an extraordinary evolutionary gamble, with no odds on how it would turn out. And for the first few million years, it seemed to make no significant difference. The creatures who opened their pelvic girdle to stand erect showed no outstanding characteristics for a very long time. During those aeons, however, they—we—developed three critical features: large brains, flexibly opposable thumbs, and fleshy buttocks. And as we became conceptual, technological and meta-sexual, we perfected our upright posture and grew taller.

Then, when these changes had crystalized, we entered a second stage of development. Somewhere around a few hundred thousand years ago we began to exhibit some unusual behavior. We began burying the dead, carving totems, drawing on the walls of caves. Our weapons and tools became more sophisticated and our social organization more cohesive. We manifested a growing consciousness of ourselves as being somehow removed from nature. We became "magical."

This period was much shorter than the first, and some ten or fifteen thousand years ago we were catapulted into our third stage of evolution as a species. This transition is generally associated with the discovery of farming, which produced surplus wealth and laid the foundations of the nation state. Subsequent changes came rapidly, including the introduction of money, the written language, and "history."

It was then that we extended verticality into time, going from a sense of cyclical recurrence into linear progression. And with that, our awe at death was transformed into a fear of death, giving rise to human sacrifice, slavery, large-scale war, and the construction of pyramids and cities—tombs for the living as well as the dead. We began not only to inhabit the vertical but to identify with it. It thrust through us and into our burgeoning civilizations. During the latter part of this period, the great myths were born, and it is not surprising that the gods, and eventually God, were pictured as being "on high." The myths reflected and supported a political order of kings and priests in which power was wielded "from above."

At this point a significant fusion occured, marking our fourth crisis of evolution. The vertical principle wedded the otherwise relatively harmless patriarchy which was dominant at the time, and somewhere around a few thousand years ago their union bore fruit. Their child was the ego, the fully individuated "I," the first symbol of which was the Tower of Babel, an attempt to split the sky.

The ego, as a function or activity, is neutral; it is just a tool. But because it was shaped by a male worldview, to the exclusion of all female qualities, it threw the brain out of balance, and soon became completely alienated from nature, turning rigid, dictatorial, toxic, simultaneously weak and brutal, imposing hierarchies onto the flux with itself at the pinnacle of any order it creates. Beyond that, its development has been uneven. The vast majority of humanity is still caught in pre-egoic and low-egoic levels. There are relatively few people who have formed within themselves a mature and fully articulated ego, much less gone beyond it.

This is the legacy of the modern age, and all attempts to heal the fragmentation seem only to have made matters worse. Certainly all political and psychological efforts have proved futile, and all spiritual programs for "slaying the ego" have been based on the same vertical consciousness and patriarchal bias which brought the ego into existence in the first place, thus producing churches and cults that are terminally hierarchical.

This is the impasse of contemporary civilization, and the argument of this book is that there is only one way to get through it, and that is to pull the plug on the vertical acceleration. The species has reached its present limits of vertical development. No transformations within the vertical order can be of any use for they are all founded on the prejudice of ascendancy, the long sweep of humanity away from the horizontal life-base.

The current world condition shows how far out on a ledge we've climbed, and suggests more than a little strongly that the edifice we've built is about to collapse. However, if we re-orient ourselves to the horizontal, we may be able to cushion the impact of the weight of the past ten thousand years now crashing to earth to serve as mulch for the new era. Also, we can cleanse ourselves of the pollution of pride accumulated during the past several million years of biological verticality. We can't go back to slouching and swinging in the trees, nor can we return to the early days of gathering and hunting, but we can learn from the negative aspects of our past evolutionary movements and allow the present change to take place without the usual confusion and violence of tectonic shifts in culture and consciousness.

More than anything else we need, as a species, to heal ourselves, and to let the Earth heal from our predations. We need to dismantle all our unnecessary and excessive vertical structures or let them decay from disuse—our hierarchies, competitions, skyscrapers, thrones, missiles. We need to get out of the way of the Way and let pure intelligence flow through unobstructed organisms and uncongested cultures. We are poised between breakthrough and breakdown, and we are exhausted from maintaining that stance. It is time that we lie down.

Lying down is the most direct way to get out of the way of the Way—a meta-meta-way. And one that has already been researched for several thousand years. As I noted earlier, the posture is found as a form of meditation and ritual in just about every Eastern tradition. At the heart of Taoism was "reclining at one's ease," preferably by a woodland stream. In hatha yoga there is *shavasana*, the corpse pose. In all communities serving a guru, *sashtang*, or full prostration, is the central symbolic act. In Islam one prays in the Masjid, the House of Prostration.

In the West, the posture appears in a number of contexts. In witchcraft it was called "grounding." Shamans entered their trance states by stretching out on the earth. The Romans feasted lying down, a practice that prefigured the *dolce far niente* of the later Italian court. Freud discovered the practice independently, but failed to see that it is the lying down and not the analysis that is the critical element. The posture is used in many schools of psychophysics, including creative visualization, the Feldenkreis method, the Alexander technique, hypnotism, massage, the isolation tank, and so on.

Coming full circle, Charlotte Selver re-introduced the "first dignity" back into Buddhism by teaching sensory awareness and deep relaxation at the Tassajara zen center, a woman thus signalling the shift from the model of Gautama seated upon a pile of straw to that of Maitreya stretched full length upon the ground.

This extensive use of the posture as a spiritual and therapeutic discipline, however, is simply the formal acknowledgement of its profound meaning in daily life. Our first year as infants is spent lying down, a significant imprinting. It is our initial posture at birth and our last at death. It is the primary medium of sex and sleep.

All of this makes the posture in its four fields extremely viable as a paradigm. It can generate models, concepts and behaviors indefinitely. Further, as the Ono-Lennon bed-piece underscored, in a world grown too noisy, busy, violent and confused, lying down can be translated into a pragmatic vision. One may even imagine that lying down has been lying in wait, as it were, for the day when the vertical thrust reached its highest and lowest levels, to bring to a halt its frenzied acceleration, and to suggest a more sane perspective to creatures who are trying to live on the figure instead of on the ground.

As I sit here writing about lying down, I notice that the day has turned to dusk. I think about sinking into a hot tub for a while, then stretching out in front of the fireplace to listen to music. After a mostly vertical day, I look forward to a horizontal evening.

I suggest that if at any time this book gets too heavy, either in actual weight or in the burden of its words, that you put both the book and yourself down. And as you do, see if you can accept the posture not only as an existential condition but as the matrix for a way of being. Consider that the awareness of a practice lies down with you, a practice that is as old as life itself, and yet, as a vehicle for the re-alignment of human intelligence and as a symbol of peaceful evolutionary transition, as new as the light of dawn.

In any case, whether or not you agree that lying down is the basis of a nascent worldview, I would at least like to convey a more subtle and far-reaching appreciation of the posture, and of postural awareness, than is ordinarily accorded it. For no matter what I write or what you read, we will all be lying down each day and each night, for the rest of our lives. Why not make it a healing and gratifying occasion?

1 Position
The Physical Field

"I am trying for a minimum of action, gradually."
—Marcel Duchamp

Each of the chapters of this book examines one of the four fields of lying down—physical, intimate, metaphoric, and mystic/absurd. This first chapter describes some of the positions in which we can lie down, the forms the body takes when it goes horizontal.

Lying down is a fluid posture. Unlike any of the sitting and yogic disciplines, there is no "wrong" way to assume the posture, nor any "incorrect" thing to do while you are in it. And since no energy is required in maintaining balance against the pull of gravity, any position is possible.

This doesn't mean that there can be no formal positions, however. Out of the very large number of variations available, we can fix certain forms as friezes, or *mudras*. For the most part, these are chosen arbitrarily, but some, like the foetal curl and the corpse pose, are archetypal. A catalog and description of all imaginable variations would require a book in itself, and here I've limited myself to delineating nine forms to use as models.

These are divided into four categories: Open to the Sky, Open to the Earth, Open to the Horizon, and Opening at Random. The text describing the positions also suggests some things to look for in each variation. However, none of this should be taken as a formula to be followed. As always, the recommendation is that you find your own way to embody the horizontal.

Open to the Sky

Corpse

This is one of the three essential postures in hatha yoga, along with sitting and the headstand. To assume the position, lie on your back, legs parted about thirty degrees, and arms also slanted about thirty degrees from the torso, hands resting palms up.

While this seems the easiest of all possible positions, teachers of yoga know that it involves subtle difficulties. Many people simply cannot let go, and even when they feel relaxed there is still tension remaining in the back of the neck, jaw, eyes, fingers, anal sphincter, inner thighs, knees, and ankles, not to mention the deep muscles such as get addressed by Rolfing. With these tensions go compulsive thinking, a general dampening of the senses, and a raft of emotional disturbances.

The suggestion of the Corpse is to die to all that. If you can relinquish all that you are holding onto, and surrender your identification not only with external objects but with memories, desires, fears, then this pose can be practiced perfectly.

This is an ambitious goal, but since dying takes a lifetime to do anyway, this position can be studied with constantly deeper understanding until, one day, it is no longer a pose but a description of your existential condition. In that moment you have transcended the practice.

DF

Circuit

Lying on your back, cross your legs at the ankles (right over left or vice versa) and lay your hands on your chest or diaphragm, fingers touching and lightly interlaced.

This is an energy-conserving and regulating position. Its purpose is to keep the body currents in a single circuit, with no "leakage." These currents can be understood as bio-electrical patterns, plasmatic streamings, or the more subtle channels such as are tapped in acupuncture.

As with all lying down positions, relaxation is a key factor, but not to the same degree as in the Corpse. The point here is not to dissolve completely but to maintain a certain psychophysical torque. The exercise involves first contacting your energy, and then letting it move freely without connecting to external involvements or internal concerns.

The mood of the Circuit is one of caring for yourself, of psychosomatic wholeness. The crossed ankles indicate closure, or organic unity, while the hands folded on the chest or diaphragm symbolize an embrace, the cradling and shielding of the heart and breath.

This position can be used for refreshing yourself at any time during the day. Ten minutes can work wonders. It is also recommended for healing practices.

Rack

Here the arms and legs are spread almost as wide as they can go. The name of this position refers to the sense it imparts of being spread-eagled on a rack. Also, from above, the body looks like an antenna receiving impulses from the sky.

The purpose of this variation is to extend the body beyond its limits, but very gently. The stretch should start from the center of the body and surge outward, slowly, reaching the limbs and fingers and toes only after all the muscles in the torso have been stretched.

The mood of the Rack is one of blissful expansion, a reaching beyond yourself, even of flying apart. Also, there is a feeling of vulnerability. The chest, belly and genital areas are open, and the arms and legs can't be used for defense. So it is necessary to contact one's fear and get past it.

The Rack can be turned into mild exercise if the arms and legs are raised slightly from the ground. Holding that position will tone all the muscles of the body. It is also a good antidote to excessive inwardness and can be used to advantage whenever you are psychically or physically cramped or hemmed in.

This position completes the Sky cycle, which begins with the total relaxation of the Corpse, moves to the tensility of the Circuit, and ends with the body stretched beyond itself and vibrating with the power of its native energy.

Open to the Earth

Arrow

Although this is the counterpart of the Corpse, its effect is quite different. The position suggests a truncated arrow. The legs, spread apart thirty degrees, are like feathers on the shaft, while the arms, also at a thirty degree angle from the torso, form the slope of the tip of the arrow leading to the head as the point. When you lie in this position for a while, there is often an actual sense of forward motion.

This is a good position for countering the frazzling effect of fragmentation in daily life. The Arrow embodies the simplicity of function, reminding us that we are being shot through creation from an invisible bow toward an invisible target, completely subject to the currents of space and time in which we find ourselves appearing.

Getting caught up in the decisions of daily life, we tend to forget that our entire life is already out of our hands. In surrendering to the larger context of cosmic helplessness, paradoxically we are empowered to be of help in the universal dilemma of existence.

This position is a reminder that what we control is infinitesimal in relation to that which controls us. In the Arrow we simply shape ourselves to go in a direction, and then let what we might call the Force spin us around and send us moving precisely where the universe wants us to go.

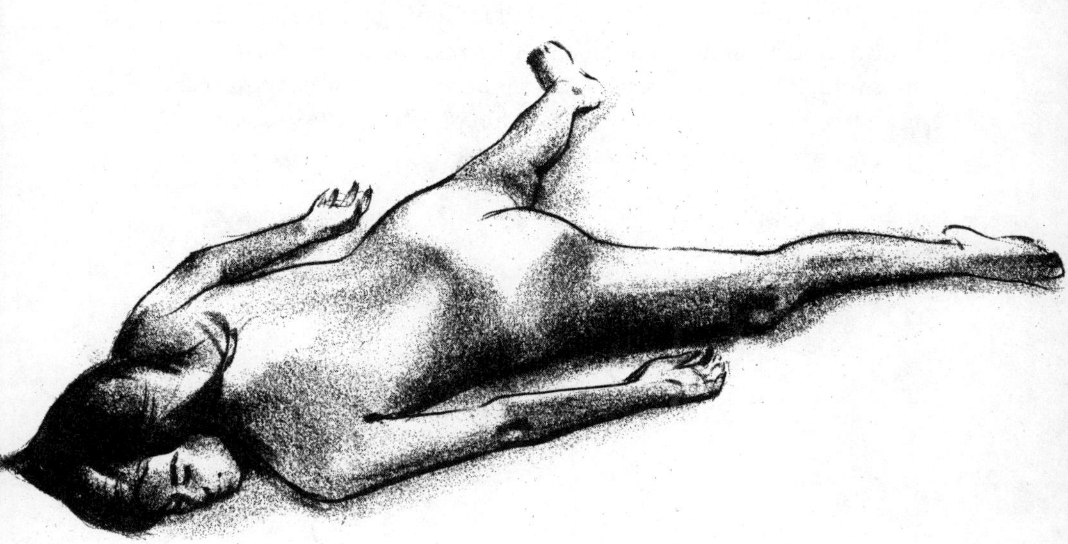

Position The Physical Field

Chip

This is the flip side of the Circuit, except that the hands do not fold under the chest or diaphragm but make a pillow for the head. The legs are crossed at the ankles. It is a position with no strong image, and in fact its value lies in its very lack of dramatic or visual associations.

Its name derives from the quality of inertness it suggests. In cybernetics, the more inert an element in a system, the more efficient it is, since it provides the least resistance or quirkiness. In human terms, however, where individuality must be included as a factor, the equivalent of being inert is not complete conformity, but becoming still and empty at the core. Thus we may remain idiosyncratic without upsetting the circuitry of evolutionary processes.

We are most efficient when we do nothing, which does not mean inactivity, but action that is pure, that has no distortion and leaves no trace. The ego does not disappear when we enter the horizontal; rather, it becomes functional: it is just the specific wiring for that chip.

This is an excellent position for letting the surface chatter of the brain settle down and feeling the more profound movements of the mind. It is also a good vantage point from which to examine personality, to see how much of one's definition of self involves essential patterns and structures, and how much is overlay and indulgence.

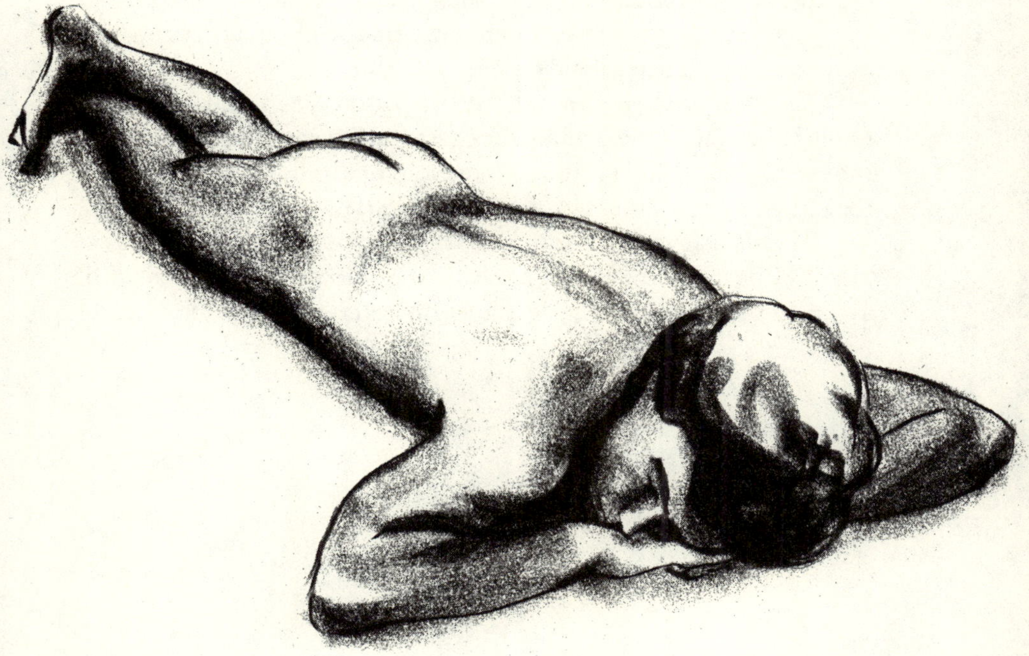

Float

This is the analog to the Rack, with arms and legs extended wide. However, instead of a feeling of expansion into the sky, here there is a sense of melting into the earth.

In this position you can easily picture yourself clinging to the huge rock of Earth that is sailing through space. A nice exercise is to orient yourself on the globe, picturing your present geographical location. Which way is your head pointed? North? East? What is the relationship between your actual physical sensations and your intellectual understanding that you are being held by gravity on an enormous spinning sphere?

The Float is good when you're intoxicated. It will allow the drink or drug to do its work as efficiently as possible. You can let waves of dizziness or disorientation pass through and over you, and let yourself be tipped into the void. No matter how wild the motion becomes, you have the very planet to hold on to.

The Float is also a pleasantly sexy position. The slight pressure of pubic bone against the ground, the chest and breasts being pressed into the earth, the belly relaxing and the thighs expanding, combined with the vulnerability of the back and buttocks, all provide a gentle arousal.

This is a favorite position of sky-divers when free-falling. And that feeling of sailing through the air is possible to attain even while lying on the ground. You, the planet, the galaxy...are all moving.

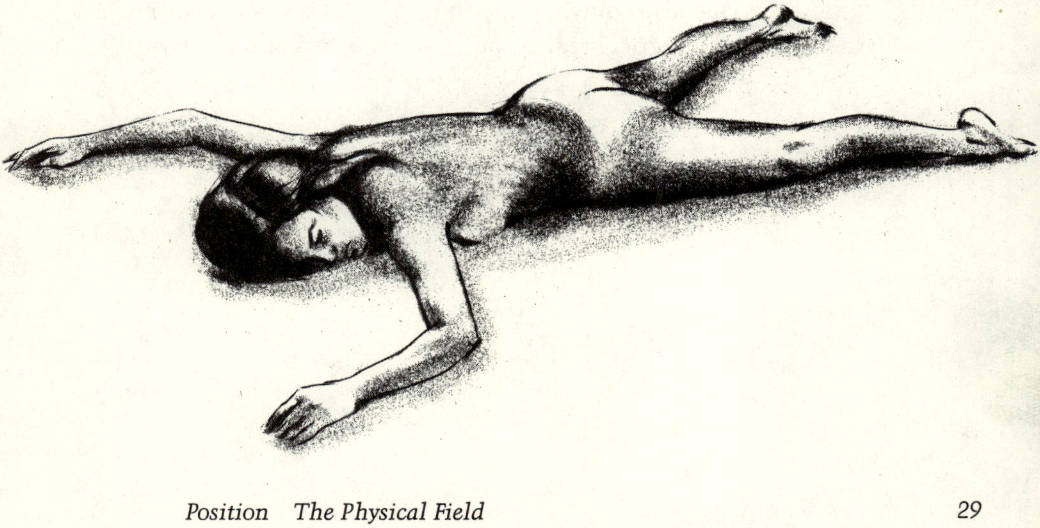

Open to the Horizon

Curl

Lying on your side, simply curl up around the navel. Legs fold at the knees and the knees come toward the chest. Arms fold at the elbows and tuck into the side. The head is brought forward with the chin reaching for the breastbone.

This position is also known as the "Foetus," because of its similarity to the natal posture, and the mood it creates. This variation involves a womblike ambiance, a sense of inwardness and slow pulses. It's possible to go into a deep regression, perhaps even rousing memories of the actual first nine months of life, of oceanic awareness, of blissful symbiosis with the mother.

However, even in the midst of such a return, one remains an adult. You are not in the womb but in the world at large, so even as you dissolve inwardly, you must define yourself externally. Your back becomes a perimeter behind which you can shield yourself, and as the center of you softens, your "shell" protects you from all hard edges and unwanted intrusions.

When you emerge from the Curl, if the movement is done slowly and with full attention, you can enjoy a sense of being reborn, of unfolding, opening, growing, expanding, and finally appearing back in the world of vertical reality, but now refreshed and cleansed.

Position The Physical Field

Lounge

This is the most active variation of lying down, even requiring the support of the head by the hand, wrist and forearm. You can lie on your right or left side, and vary the position of the legs and of the free arm.

There is something intrinsically elegant about this position, an attitude that is manifested very clearly in representations of the Reclining Buddha, descriptions of Greek symposia, and in such paintings as the Odalisque. And when you take this position, it is a good idea to bring that sense of graceful power and beauty to it.

The Lounge lets you rise to the greatest height possible while still lying down. There is a hint of verticality in the position, which provides an interesting ambiguity, even an irony, to the mood of the basic posture. Also, when you are lying in this way, you can see the horizon, so the physical horizontality is amplified by visual horizontality.

This is a good position for thinking and reflecting. Also, with couples and small groups, it's a delightful position for conversation.

Opening at Random

Doll

This is the most commodious of all the positions, the one in which you need take no heed at all for how the body spreads out or bunches up. You can lie face down, face up, on your side, with the torso twisted, and the limbs and head assuming any attitude whatsoever.

This position allows for the most graceful and the most awkward variations. You can stretch in visually elegant ways, or scrunch up in ugly contortions. You can assume poses that are pleasurable or even painful, depending on what you want to explore at any given time.

One possible exercise involves body imagery. As you lie in a given form, see what images that position inspires. Then let the image change, and shift the position of the body in relation to that change. Go from position to image and back again, even making sounds out of the movements. These dramatic extensions can be extremely cathartic, including laughing and crying, and spark insights into personality and character structure.

The Doll is a position of experimental freedom. Through it you can embody your most profound conceptions and most fearful notions. You can empower the imagination with all the force of racial memory. You can know what it's like to cut through the water like a dolphin, glide through the air like a hawk, lie broken on a rock after a fall from a cliff, be made love to, be crucified, be one with every level, dimension, texture and form in the universe.

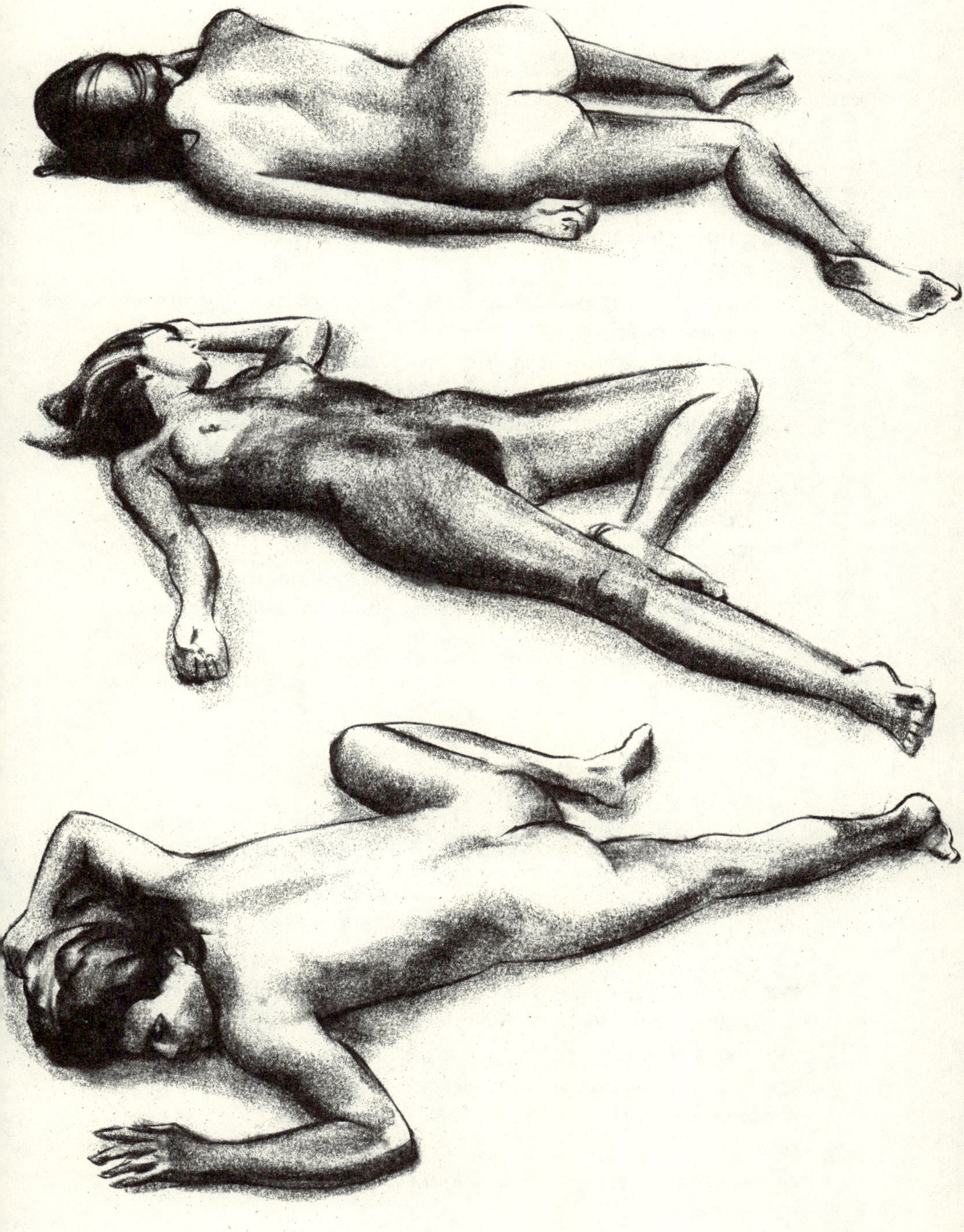

Pronepourri

This is a horizontal analog to the Sun Worship exercise, the twelve-step stretching movement that's used as a warm-up in hatha yoga. Pronepourri involves assuming all of the nine positions just described, and going through them as a continuous form. This can be done in one of two ways.

In the first approach, you stay in the first position, the Corpse, for as long as you like—three to five minutes is a good duration—and then remain in each subsequent position for the same length of time, taking the positions in order. The idea here is to make a tour of the posture, to lie down as flexibly and thoroughly as possible. This is an excellent exercise in itself, and a perfect prelude to a regular session or class that focuses on a single position.

In the second variation, you go into the positions in any order and stay in each for however long you want. The purpose here is to find the essential mood of each position, to uncover "what it wants to tell you." This can serve as a form of meditation, and invariably provides a fine psychophysical cleansing and tune-up.

"Body in the Dark"

This provocative phrase is a chapter title in *Sleep Positions: The Night Language of the Body*, by Samuel Dunkell, M.D. The work, one of the rare few dealing with horizontal body language, studies the subject in terms of psychological interpretation, examining nighttime positions as indicators of what they reveal about daytime personality.

"Throughout our lives, in sickness and health, living alone or in a couple relationship, our sleep world behavior unfolds the story of our particular existence. It reflects every turn of the plot, every crisis, every change. In fact, as we have seen, our sleep world behavior sometimes reveals our true feelings about a situation or relationship before we have even begun to face those feelings in the waking world.

"It is a rich world, this dark universe in which we spend on the average over twenty years of our lifetimes. It is unlike any other world of which we have knowledge, with its own special laws. The orientations that govern our waking lives are transformed in sleep. The dimensions of space and time are revealed in special ways in sleep consciousness. Our sense of ourselves expands to take in the myriad facets of our universe of experience. We can be a rock, an animal, another person—all in rapid succession. In the sleep world we partake richly of the vast possibilities of existence and of a sense of the cosmos.

"We lie upon our stomach, our side, our back. Our legs may be spread wide or tightly clenched, our arms extended or pulled in close about our body. And these positions in the darkness tell the true story of our changing lives, of how we feel about ourselves, our day world and the important people in it. In the night our bodies move toward or withdraw from our sleep partners in a vivid explication of our feelings about our husbands, wives and lovers."

Whether or not one agrees entirely with the specific analyses of various sleep positions, the book is practically unique in its extension of the vocabulary of lying down variations. Dr. Dunkell lists two dozen positions, including such exotic poses as Chain Gang, Sandwich, Flamingo, Sphinx, Monkey, Cyclops, Dutch Wife, Barrymore, and Swastika.

2 Exploration
The Intimate Field

"This breath going through me is all that matters."

—Veronica Vera

The second field of lying down is that of intimacy. As this word is ordinarily used, it means close familiarity with a person, or is a euphemism for sexual relations. Here, however, I'm using it more in terms of its intrinsic meaning, which refers to the inner or inmost nature of a thing, its fundamental character. In the human realm, this translates into intuitive knowing, a direct gnosis of the heart.

Intimacy, of course, is not restricted to any physical posture. You can be inside your breath while standing and in profound communion with nature while seated. But it is in the horizontal that intimacy finds its most natural and congenial ground. Lying down is the most nourishing matrix for the emergence of intimacy.

This is not, however, an automatic or mechanical correlation. One can lie down and remain guarded, distant, alienated. So, assuming the physical posture is not necessarily sufficient to bring about the radical shift in consciousness that occurs when the mood of intimacy is awakened. Nor are there any techniques for "producing" intimacy. It may be courted, but not commanded. It flowers, or opens before us, in its own time and way, and our only choice is whether or not to sniff its fragrance, whether or not to leap into its abyss.

The only thing that is possible, while lying down, is to attempt to undermine all the vertical structures, tensions and obstructions that prevent the easy movement of supple intelligence through the organism and self-sense. With the removal or dissolution of character armor, hierarchical thought patterns, distortions in perception, and all the other neurotic and psychotic mechanisms bequeathed to us by our civilization, sensitivity to our true nature is uncovered. At that moment it is possible to know reality directly, simply. This is the true process of education.

As with everything else having to do with lying down, though, each simplicity contains a great complexity. Already, in the physical field, we found that there were many positions possible, each with its own set of images and feelings. And in this second field, there are easily as many aspects, categories and states to consider.

This proliferation stems from the fact that lying down does not favor a fixed method, such as counting the breaths or keeping the mind concentrated. Its basic operational mode is the free-form drift of sentience, or awareness, through all the levels, dimensions, structures and processes of what Nina Bull calls "the body and its mind."

For example, you might start off by watching the breath, then shift into sensing the pull of gravity in your muscles. You might drift into a hypnogogic trance or reverie, and return from that to a fantasy. You might sleep, or daydream, or find your senses suddenly hyper-alert, registering light and sound and smell with extraordinary sharpness. You might lapse into ordinary linear thinking, or turn your attention to some problem of daily life, or be taken by an idea of profound significance. From there you could become aware of heartbeat and circulation, which might lead to musings on mortality, which might rouse fear or longing. You might take a stroll down memory lane into your childhood, or find yourself suddenly out-of-body and whizzing past Pluto. You might become sexually aroused, or contact some part of the body-mind that needs healing.

There are two ways to manage this profusion of experiential richness. One is to establish purpose and parameters before you begin. Purposes include relaxation, introspection, healing, pleasure, et al., and once that is established you can decide upon "set and setting," i.e., indoors or in nature; in silence or with music; on your own or with a teacher; being by yourself or with a partner or in a group; remaining still and quiet or permitting yourself movement and sound; being nude or clothed; using psychotropics or fasting or making no special adjustment for the session; going variously into trance, reverie, sleep, dreams, hypnosis, or remaining in ordinary waking consciousness.

The second way is to place no limits at all on the form of the session. Here you only enter the special quality of awareness that is activated when you lie down, and stay with that through all the changes, seeing where they lead. This awareness is impossible to describe, but when you feel it, it is unmistakable. There is a palpable shift in one's sense of being. This is the whole-body recognition that you have left the vertical, with its necessary struggles of maintaining balance against gravity and all the games of survival. It is the homey feeling of the horizontal, a sense of existential ease.

When this awareness is activated, you can move with it through all the aspects and compartments of the organism and self-sense without being pulled into identifying with any of them. You transcend everything by pervading everything. This is the discipline of empathic freedom.

However, no matter how healing and/or enjoyable all of that is, it is simply the prelude to the real awakening in this realm of exploration, which is the awakening of intimacy, or supple intelligence, or the gnosis of the heart. When that occurs, then all the rest—methods and insights—fall away. For then there is only consciousness as life aware of its own mystery.

In this state you can rise and address your vertical existence with perfect clarity, seeing through the frenzied projects, stressful social organizations, pointless momentum. And you can bring with you the freshness of the horizontal perspective in which lying down and intimacy are the first two fields.

Session

Lying down is taught as an aspect of so many disciplines and by so many teachers in so many ways that it is impossible to describe anything like a standard or average class. Each approach has its own emphasis, techniques, goals.

All I can do in this chapter is provide an example of one of my own sessions, and even here I have to be extremely selective. The classes I conduct with individuals are too idiosyncratic to be of use in a book intended as an overview. In those classes everything I do is in relation to the state of the person I'm working with, and picking any one at random would give a skewed impression of the process.

With groups, I do work within a structure, but here too there is a great deal of variation. First of all, I must leave myself room to move within the texture and mood of the particular gathering of people. Then there is the question of the stated purpose of the group, the age and gender of its members, their experience in this practice, and so forth. Lying down has been used for everything from a form of communal prayer to preparation for an orgy, and it has been practiced by everyone from six-year-olds in Montessori schools to octogenarians in hospices. Obviously, there can be no single "correct" content or structure for a group session.

However, it is possible to factor out some basic elements that serve as a foundation for all variations. In terms of my own experience, I've divided these fundamentals into two categories, or classes. By and large, these constitute a comprehensive introduction to this field of exploration.

The first class is called Impression. It is conducted in the Corpse or the Arrow, and the basic elements are relaxation, breathing and the senses.

The second class is called Expression. It begins in the Curl and extends into the Doll. The basic elements are memory and fantasy, and participants are encouraged to make sounds, laugh, cry, babble, etc.

Using these as a matrix, it is possible to move in any direction. Other classes can focus on any of the purposes already mentioned, or amplify any of the processes. One can spend an hour just on breathing, or one of the senses, or use the advantages of the horizontal to delve into life problems, and so forth. Ultimately, one can simply play.

In translating these classes into writing, a great deal is lost. So much of what happens is found in the silences, gaps, vibrations, telepathic communication, and presence that lie behind the words, as well as in the tone of voice of the teacher. But in order to give at least some feel for the process, I'm including a transcript of a session in the first type of fundamental class, Impression. I offer it not only as an example, but as a compass for your own sessions of exploration.

For this session, we'll use the Corpse position. We'll be working clothed, but take your shoes and socks off, loosen belts, brassieres, or anything else that is constricting. Then take a few minutes to wander about in the space and find a spot that seems right for you.

Before lying down, do a few stretches—up, down, to the side. Roll your head around on your neck. Wring out your fingers. Yawn. Sway or jog or dance in place. The point is to get your circulation going and the muscles tingling a bit so that when you lie down it is not like a dead weight, but with a sense of liveliness and alertness.

Then take the position. On your back, legs apart about thirty degrees, arms thirty degrees from your sides. Now for a minute or two, do nothing but settle down. Let your heartbeat slow down. Take several deep breaths. You will be lying in this position for forty minutes, so begin to inhabit it, make peace with it.

If you need to, adjust your position slightly. If there's anywhere you feel discomfort, move that part. Shift your arms and legs, or roll your head from side to side and see whether your neck is cramped. Raise the shoulders a bit and let them fall. Take an inventory of your body and find any spots where there is tension, and let that go as much as you can. Be aware that there are deep muscles that are not so easily contacted, and in just that awareness alone see if you can feel an overall relaxation of the musculature.

After adjusting for any physical discomfort, look for resistance of a psychological kind. Some of you may feel a sense of claustrophobia, as though you were locked into a tight place. You may suddenly feel that all this is silly. Or there may be a sense of being threatened in some way.

All kinds of attitudes and defenses we carry around in our vertical lives fall away when we go horizontal, and there is often an impulse to protect all that, which we identify with our sense of self. You may have fears of being exposed to yourself in painful ways, or of being revealed to others in the room. Locate any feelings of resistance in the body and mind, and simply be aware of them. Keep letting go and breathing, and they will melt away.

Take a deep breath, let your weight sink into the floor, and put your attention into the physical condition. You are lying down, supported by the ground, and in a safe place. Nothing here will harm you. You may have thoughts and feelings that touch you deeply and you may experience those as dangerous or unsettling, but they are only signs of energy flowing through you in a freer way.

You have been lying down for several minutes now, and you may be beginning to feel the first changes in your sense of time. As we go more deeply into this posture and process, there may be moments when it seems as though it is going on forever, and when it is over it may seem to have gone by in a flash. Both are true. As is the chronological measure of forty minutes. I am taking responsibility for the clock, so you can allow yourself to get lost in your subjective sense of time, and timelessness.

Most of you already have your eyes closed and I'll ask the rest of you to do that also. This is both to prepare for a specific relaxation exercise for the eyes, and to allow you to concentrate on other forms of perception beside sight, which is overused in our culture, to the detriment of the other senses.

The central channel during this session is hearing. The connecting thread through these forty minutes is the sound of my voice in your ears. The content of what I say will vary, to include instructions, suggestions, images, areas of emphasis, but the basic medium of transmission is sound itself. And for optimum effectiveness, it's best if you forget about me as any kind of presence or person, and have my voice be disembodied, as though it were one of the voices you hear all the time in your mind as you go through the day. Like those other voices, mine will fade in and out of your attention. At times you will forget about it altogether, like background noise. At other times it will seem to stimulate the very core of your brain. But in any case, just ride it like a magic carpet and see where it takes you.

Now, with all this—coming to terms with any resistance you may be feeling, forgetting about clock time, tuning into my voice—turn your attention to your sense of touch. Specifically, become aware of where your heels, calves, thighs, buttocks, spine, shoulders, arms and head touch the floor.

Touch has several aspects. The first is simply pressure. Your body, being gently, consistently, pervasively pulled by gravity, exerts pressure against the floor. Feel that. Is it comfortable or uncomfortable, hard or soft? See if you can get a sense of gravity itself. It is a force you have to balance or struggle against when you are vertical, but now it is not only not antagonistic, but is soothing, nurturing. See if you can surrender to gravity, as it were, have it massage you, embrace you. Let it enfold you. Yield to it. Let go.

Another aspect of touch is texture. What is the texture of the surface you're lying on, smooth or rough, firm or soft? Or is it ambiguous? Are there places where you can't tell what the texture is? Can you flex your imagination a bit and erase the sense of distinction between you and the ground? Can you feel yourself merging with the floor?

The front of your body, not touching the ground, is touching the air and your clothing. Can you feel the air against your skin? Is it cool or warm, still or moving? The air too has pressure. It too is being pulled by gravity. The air that envelops you is being pulled toward the Earth over and around and against your body.

There is not the smallest area of your skin that is not touching something—the ground, the air, your clothing. If you are sensitive enough, the entire envelope of your skin will come alive with touch. And as you sink into the ground, and breathe, become only the sense of touch for a moment. Touch the earth, the cloth, the air.

Now turn your attention to your breathing. Don't do anything special, just watch it. And you'll find that by becoming aware of the breath, it will change. Let it happen. Get a sense of the importance of breath, how primary it is. Rest in it for a while.

You might try, on your next exhalation, letting the air out with a hissing sound, like steam escaping from a valve. Make the sound until there is absolutely no breath left, until you have emptied your lungs completely, and then relax and let the next breath take itself. The secret of proper breathing is in the full exhalation; the inhalation then follows automatically.

There's an interesting way of watching the breath, which is to follow its movement into your body. Feel its movement as it goes past your nostrils, into the nose, past the sinus cavity, into the throat. See how far back you can actually feel the breath going down, how far the sensation takes you. As you go down into it, there comes a point where you're not sure whether you're feeling it or not, whether or not you're imagining it. When you find that point, then let *it* start moving further down. You follow the breath past the nostrils, into the nose, into the throat, until you reach a point where you're not sure what you're feeling is sensation or imagination, and then move along that edge down into your lungs, feeling the ribcage lifting, the diaphragm distending.

If you can stay at the edge between sensation and imagination, then each breath can be a whole-body experience. You can feel the air being absorbed by the lungs, the oxygen entering the blood-stream, circulating, purifying. You can get lost in the movement of the breath and its chemistry through the body. And each time you exhale, you can feel the toxins being expelled. Don't try to gauge how much is sensation and how much imagination, just go for the feel of it.

Now see if you can link the movement of the breath with the pressure of your body being pulled by gravity into the ground. Make the connection between relaxation and breathing. Each time you exhale, you get heavier; and when you inhale, you get lighter, like a balloon being filled. Each time you exhale, let yourself sink more deeply into the ground, and each time you inhale, get the sense of floating. To do this, you have to progressively relax all your muscles.

Let your feet fall apart and let the thighs relax. See if you can extend the breathing into your lungs. Use the edge beween sensation and imagination to follow the breath down to your feet, through the thighs, knees, calves, ankles. As the breath fills you, pay attention not only to the physiology of breathing, but to the overall sense of breath filling you. Let the body be hollow and let the breath be a stream with many waterfalls spilling into it. Feel the breath not only in the nose and chest and belly, but in the legs, the shoulders and arms and fingers, in the skull and brain. Let relaxation be not just letting the muscles go slack, but an inner melting and dissolving. Let it be voluptuous.

For a few minutes reduce all your sensation, perception and feeling to those two processes—the breath filling you, and the pull of gravity drawing you more and more deeply into the earth, while your muscles continue to let go.

Now turn your attention to your eyes. First, realize how large a part sight plays in your life. Become aware not of the objects of sight but of the meaning of sight. Consider how much of what you are depends on seeing. Get into that awareness, for what we will be doing is relaxing some of that, disengaging from the identification with sight.

The basic feeling we're aiming at is not that of being blind, but of being sightless, like a tree. This is a state of sentience where there is feeling and awareness, but no sight. Try to get a sense of what it would be like to have consciousness of yourself and the world around you, but never to have seen it.

Become aware of the eyeballs themselves, the actual physical organs. Can you get a sense of their peculiar consistency, the many muscles, the intricacy of the blood vessels, the connections to the brain, all of that? Realize what an extraordinarily complex system they contain and represent, all of the millions of discriminations of which they are capable, and the way they transmit impulses to the brain which then translates those signals into images, colors and meanings.

Try to grasp the difference between the significance of the eyes and what they see, and their simple physical makeup—the two globes of hardened jelly. Become aware of the head in which they rest, the bony structure, the actual skull. Understand the eyes in their starkest reality—two small spheres of soft tissue in a bony skull. Make contact with the way they are cradled in their sockets, their fragility. Forget all about them as organs of sight, and feel them only as organs. Pull back from the world they open onto and go inside the body where there is no sight but only awareness.

Even after the eyes are closed the muscles continue to work out all kinds of conditioned patterns. And what we're doing now is resting the eyes in a way that you can't even when you're asleep. Do this by becoming aware of them as organs, forgetting what they do. Let the eyes sink into their bony cups and let all the perceptions, relationships and meanings that are connected with sight let go also.

As you do this, continue to give yourself over to breathing, that delicious feeling of doing nothing but being filled with breath. Be like a jellyfish being rocked back and forth by the tide, rising with each wave of breath coming in, and sinking with each wave of breath going out. Feel yourself entering a rhythmic sense of time, a natural cycle.

With all of this there will be various sensations in the body—feelings, memories, fantasies, thoughts. Let them happen. Wherever you contact a feeling, sensation, memory, idea, let the breath go there and amplify it. Feed it, let it blossom, see what it is. It could be pleasurable or painful. It could be the sudden emergence of a fear you've been hiding from, or a desire you don't want to face. Or just something in the body that's wanted to make itself felt for a while. Perhaps a blissful expansion, or a contraction in the psyche.

As you let the structures of the mind relax by relaxing the eyes, and are cleansed by the movement of the breath, and massaged by the pull of gravity, you are simplified, and can observe what happens inside you much more clearly. And when you've experienced whatever feeling or thought has arisen, let it pass, let them all pass, like clouds sailing above the horizon.

Realize now that none of this is you. Each of you is experiencing a variation of the same state. When we enter the state of awareness, we are one mind, one body, one feeling, one thought. This state doesn't belong to anyone, it doesn't define anyone.

What makes you unique is the individuality of viewpoint you embody. Although the process is universal, no one in the universe can know it from precisely the angle you do, and with the details you bring to your perception. What you are is simply this—a unique viewpoint in the universal process.

See the thoughts, ideas, fantasies, fears, ecstasies, memories, and discover what the relationship is between all that and the breath going in and out, the pull of gravity. What is your personality in relation to this body of sensitivity lying on the ground? And how much of what is happening now can be awake when you sit up and stand up and begin walking and talking and being the person you are in the world?

Now see if you can slip off even that consciousness and into a pure reverie, where everything becomes very, very still. A good image to use is that of a log in a forest that's been lying there for a hundred years, decomposing and becoming overgrown with moss. Get a sense of the timelessness and stillness of the deep woods where no human goes. There is no sight or movement now, no thought, no feeling, but only awareness, life knowing itself simply. Enter that. And if anything arises, let it sink and be absorbed into the soft forest floor. Finally, let even your awareness dissolve and drift into conscious oblivion.

Then, gradually come back from that to the real stillness of your body, the actual silence of the room. Return to your ordinary awareness, the pull of gravity, the breath sweeping in and out, your eyes sunk deep in their nests. Let the traces of the trance lift off slowly, like morning mist.

And whenever you are ready, let your eyes open. But when they open, don't reach out to look. Let your eyes stay as relaxed as they have been and let the light come into them without structuring it and giving it shape and meaning. Be receptive. Look upon the world with the wonder of a child.

Lie as long as you want to until you feel like stirring. Then, when you do move, do it slowly. First the fingers and toes, and then the hands and feet, and then the arms and legs, and finally the torso and head. Bring the whole body into the movement slowly. Slowly.

Some Responses by Participants

"For a while I felt like a child. I really went back to when I was about eight years old. And from there it went into pure illusion. There was a feeling that the air was going right through me. And then I disappeared."

"It was difficult at first. I felt tense. And then felt guilty for feeling tense. I told myself, 'How can you do this to yourself, let yourself feel so bad with everything so tight?' Then I told myself to let go, and the feeling of relief was incredible."

"It got very sexy. I felt open, vulnerable, sensitive. Every breath was like an orgasm."

"It was wonderful, it was paradise. At the very end I felt like I was attached to such a little, tiny thread. I wasn't afraid, but there was nothing left except that thread. I think it was your voice. And I felt that if I could let go of that, I could die."

Special States

While working with a teacher has certain advantages, there are situations in which lying down alone yields uniquely beautiful and meaningful experience.

One is simply lying down in nature in a place where no human artifacts can be seen. This could be in deep woods, on a beach, in the desert. The absence of people and manufactured objects has an extraordinarily liberating effect which, when coupled with lying down, is profoundly refreshing for both body and psyche. Evelyn Eaton, a Canadian woman of Micmac and Malisite descent, who has been initiated into the Sweat Lodge and Sacred Pipe rituals of the Paiute and Arapaho medicine people, describes this practice: "Then I went toward the mountains. There was a tangle of willows by a stream where I sometimes went when I needed renewal. A good way to refill depletion is to lie on the earth, stretched out, head to the north. Normally I lie on my back, face to the sky. In sorrow, defeat or emergency, I lie on my stomach, face to the earth. It works. The healing currents flow." Once one feels the power of the process, it's possible to make contact in heavily built-up environments as well. "Even in cities," she continues, "walled away by concrete slabs and thick cement between us and the sky, the currents still flow, slower, fainter, and it may take longer to heal us, but it works."

A second situation involves totally opposite conditions—lying in an isolation tank. Here one is as solitary as in nature, but is surrounded completely by a human artifact. This extremely artificial environment structures effects that have been examined and documented in detail by John Lilly, and taken to their transcendental sci-fi conclusions in Paddy Chayevsky's *Altered States.* Lying weightless in absolute darkness and silence is, for many people, a terrifying experience initially, but very soon opens into realms that transform those who enter them.

A third circumstance is more accessible than these two, and that is simply lying in bed alone for long periods of time. The length of such an experiment depends on one's schedule, of course, but giving over a weekend should be possible for everyone. It is a wondrously cleansing process in which seemingly thorny complexities are often clarified or even dissolved just by letting them go hang for a while. In *The Summing Up*, W. Somerset Maugham describes an enforced period of being confined to bed in a sanitarium because of tuberculosis.

"For the next two years I led an invalid life. I had a grand time. I discovered for the first time in my life how very delightful it is to lie in bed. It is astonishing how varied life can be when you stay in bed all day and how much you find to do. I delighted in the privacy of my room with the immense windows open to the starry winter night. It gave me a delicious sense of security, aloofness, and freedom. The silence was enchanting. Infinite space seemed to enter it, and my spirit, alone with the stars, seemed capable of any adventure. My imagination was never more nimble; it was like a barque under press of sail scudding before the breeze. The monotonous day, whose only excitement was the books I read and my reflections, passed with inconceivable rapidity. I left my bed with a pang."

A fourth situation is the most accessible of all, and that is the bath. Lying in a tub of warm water is universally considered a pleasurable and healing experience, but in the following passage Paul Valéry captures its poetry.

"In the pure and shining sarcophagus, sweet water reposes, warm and perfect, bride of the body's form. Light and free, the nude settles down and is assuaged. Everything is easy in the fluid... Here man has laid his stature down; here he has glided his whole length into which his height has changed; he stretches out until he reaches the limit of his own extent; he feel commensurate with his feeling of his capacity to relax. Blissfully he changes his point of support; one finger holds him, lifts him; and his floating powers, half melted in the calm mass of the water, dream of algae and angels. The weight of the blessedly bathed flesh is almost imperceptible; the heat of the blood being hardly different from that of the contiguous water, the blood spreads out under every inch of skin. The living body is hardly distinguishable from that formless body whose substance is constantly taking its place. A person is mixed with the indefinite plenitude that surrounds him; someone feels himself being dissolved. The whole body is now only a pleasant dream that the mind vaguely entertains. The sweet moment looks at itself and through the window of the water sees limpid limbs. What is looking at and what is speaking to oneself is astonished at the size and symmetry of the limbs over which it rises; and the thinking head is amused by a foot that has appeared far away and that obeys it as if by magic... The will itself and the general freedom of the being are calmed in the comfort of the water. Perhaps in the vapid and vaporous air there is a perfume whose complex flower stirs memories, caresses or colors the indistinct desires of the naked being. Eyes waver and close. Without contacts, duration disappears. The mind opens its veins in a dream."

A Zen Addendum

The following are excerpts from a talk by Taizan Maezumi, Roshi, published in "The Ten Directions," the newspaper of the Zen Center of Los Angeles and the Kuroda Institute.

"Quite often in my talks I use the word 'intimate.' Be intimate with yourself! That's what the Dharma transmission is—really being intimate.

"The real secret teaching is no secret at all: being intimate. The best teaching, the best truth is no place other than within you. No one outside of you can do anything with it. Even you can't do anything with it, except to be just as you are. Just be so. If anything is secret, your very life itself is secret. Buddha realized that being intimate is actually the key. That is what he transmitted to Mahakasyapa, and that is what Mahakasyapa realized. Generation after generation, that is what has been handed down. It is our very life itself.

"*Mitsugo*, by Dogen Zenji, is usually translated as *The Secret Teaching*. But literally translated, *mitsu* means 'intimate,' and *go* is 'the word.' In this case, 'secret' means the same as 'intimate.' Since the teaching is so intimate, it becomes secret. Yet is it not really secret when you understand that intimacy. One section reads, 'The secret teaching was not only possessed by Shakyamuni but by all the Buddhas and Patriarchs. Therefore, anyone who is a teacher to the world must surely possess a secret teaching.' This secret teaching, as I just mentioned, is the same as intimacy—intimate word, intimate understanding, or being intimate.

"In a famous koan, the Second Patriarch is asked by Bodhidharma, 'Bring me your mind which is not at ease.' After spending eight years looking for it, the Second Patriarch finally reached realization and then said, 'The mind is ungraspable.' He is talking about the same thing, expressing that state of being intimate. Really being with yourself with no division between self and other. Then everything becomes nothing but you. Nothing could be more intimate than that. That is Buddha's teaching.

"In your daily life, please just accept and appreciate yourself as you are and your life as it is. I want you to take good advantage of whatever chance you have to become a really intimate being. Don't hurt yourself. Just really be intimate with yourself."

3 Translation
The Metaphoric Field

"When in doubt, get horizontal."

—*caption on a picture postcard*

The third field of lying down is the metaphoric, in which the idea of horizontality is abstracted from the posture and its mood, and applied conceptually and culturally to daily life. This involves uncovering or underlining all the vertical assumptions in human thought and behavior, and translating each attitude and artifact into the form it might take were it to emerge from a horizontal understanding instead. Ultimately this would mean dismantling most of our civilization, as well as dissolving the personality types and character structures it has produced and which now sustain it.

It seems a forbidding project, but fortunately it is not something that anyone has to "do." It is all happening already anyway. The crisis of the modern world is actually nothing other than the first global intimation of the coming collapse of the vertical mythos, logos and technos. The crash is far more profound than the decline and fall of any particular nation or people; it is the destruction of the essential bias that has defined us as a species from our very inception. We are witnessing the end of the vertical joyride, with all its splendors and horrors.

Except that society is a conservative and even reactionary beast, this would be cause for elation. It would seem that the best thing imaginable right now would be for us to extricate ourselves from the vertical acceleration, and enter an epoch of horizontality in which ease and slowtime would reign. There ought to be a great relief in contemplating an end to "progress," for isn't there a hunger in all of us to enjoy our lives instead of having them used as fodder for some deranged system and machine?

This book is being written with the certainty that there are millions of people who have been thinking and feeling this way, and tens and hundreds of millions who can't allow themselves such a vision because of fear—of letting go their vertical identities, or that their government or economy would punish such waywardness.

However, whether this idea is celebrated or suppressed, the decay of the vertical hegemony goes on. The only question is whether its fall will take the form of a cataclysm, or whether we will cooperate with the inevitable by moving gracefully, instead of violently, in the evolutionary direction we'll have to travel anyway.

What we are doing now is not working well at all. With increasing population, vastly unequal distribution of goods, nuclear weapons, industrial pollution, and the near bankruptcy of the world economy and other symbol systems, the final curtain is ready to drop on the vertical drama. And there is widespread tacit awareness of this. With the explosion of the first atomic bomb, we were collectively sobered by that stark vertical artifact, the column of fire—biblical in its import—rising into a mushroom cloud. From that moment on, with growing suspicion that we were close to destroying ourselves and most of the life on the planet, we hurled ourselves into a frenzy of vertical excitement and sent rockets out of the gravitational pull of the Earth, a glorious and desperate gesture by the vertical mind at the brink of suicide.

As I write this, in late 1983, all of this activity seems to be in abeyance. Space probes have become perfunctory, or militarily secret. And nuclear weapons have proliferated to the degree that even the dimmest minds can perceive that using them is ultimate madness. So there is a momentary slackening of speed and an uneasy balance of forces. And the question hanging in the air is: is this the point where the species turns its trip around, or is this the lull before the big storm?

If it is the latter, then this book is simply one more object in the rubble. But if there is a miracle of transformation, then, I am arguing, the medium of that change will be lying down. Which, in its third field, means assuming horizontality as our primary metaphor, our "paradigm." And I am arguing for this shift without knowing whether this would be a period of rest and integration before another cycle of vertical movement, or whether this will prove a permanent alteration, like that of the dolphins returning to the sea, those uncanny creatures who seem to be trying to communicate the horizontal virtues of simplicity and peacefulness to us.

But whatever the long-range or ultimate outcome, in its immediate effect lying down constitutes the cutting edge of conceptual and cultural metamorphosis. Since it undermines the vertical monolith in whatever form it appears—under whatever flag, icon or ideology—it is the most radically subversive idea in the world. Lying down is the real revolution of which all the "revolutions" of the past several thousand years have been distorted foreshadowings.

In this chapter I've noted some instances of the horizontal shift as it has already begun to take place in various sections of society, and have also made suggestions for such a translation in other areas. I don't put any of this forth as a proof of any sort, but rather offer it as a lens through which to see oneself and one's civilization from a different angle and dimension.

At the end of the chapter I've added a collection of quotes from various classical and contemporary sources, all of which reflect on one or another of the fields of lying down.

There is a striking description of the vertical to horizontal shift in organizational power structures in John Naisbit's *Megatrends*. He writes about networking.

"Networks are people talking to each other. Networks exist to foster self-help, exchange information, change society, improve productivity and work life, and to share resources. They are structured to transmit information in a way that is quicker, more high touch and more energy-efficient than any other process we know. Networks offer what bureaucracies never can deliver—the horizontal link. The vertical to horizontal power shift that networks bring about will be enormously liberating for individuals. Hierarchies promote moving up and getting ahead, producing stress, tension and anxiety. Networking empowers the individual, and people in networks tend to nurture one another."

The equation of verticality with aggression and alienation is clear, as is that of horizontality with more humanly positive qualities. Such a shift is much to be desired, but the question to be considered is to what degree this change will be resisted by those who control hierarchical power bases in state, church, business and education.

Edward de Bono's "lateral thinking" is a casebook example of the horizontal shift. Whatever the value of the school and system he has founded on the basis of the concept, the idea itself illuminates the differences between relaxed and rigid thinking styles. Among the ten basic differences he lists between lateral and vertical thinking are the following: "Vertical thinking concentrates, and excludes what is 'irrelevant,' while lateral thinking welcomes chance intrusion; vertical thinking moves only if there is a direction in which to move, while lateral thinking moves in order to generate a direction; with vertical thinking one has to be 'correct' at every step, while with lateral thinking information is used for *effect*—it may be useful to be 'wrong' at an intermediate step in order to obtain a better arrangement of information."

This sense of moving about sideways in thought was given an unintentional amplification by, of all people, Queen Elizabeth. During an audience, while de Bono was describing his system, Prince Charles walked into the room and, hearing the phrase "lateral thinking," asked what it meant. "I haven't had it all explained yet," said the Queen, "But from what I gather it's the sort of thinking one does while lying down."

History as his story, not hers, or anyone or anything else's. Not surprising since the idea of history was born at about the time that the principle of verticality was deified by the partriarchy and used as the justification for male superiority. History begins as a chronicle of kings, priests and generals, with their tyrannies, hierarchies and wars. And the subsequent study of history has involved a vertical digging into time and into the earth, thus strenghening the prejudice that history progresses linearly and upward.

But life is a web on the surface of the planet. What happens anywhere affects what's happening everywhere, and the human story is only a very small aspect of the whole. Choosing one culture, one nation, one gender, one race, or even one species to study as an isolated and privileged occurrence reflects monumental tunnel vision.

There are not many historians, unfortunately, who can see beyond their specialty. Werner Stein's *Kulturfahrplan*, rendered by Bernard Grun into *The Timetables of History: a Horizontal Linkage of People and Events*, moves in the right direction, but its categorizations are crude and it covers extremely limited ground. And metahistorians like William Irwin Thompson don't seem to have much effect on the academic establishment.

In its vertical mode, history can do nothing but inculcate divisiveness and narrowmindedness. In the horizontal paradigm, history is dissolved and replaced by the more universal and accurate field of ecology.

Vertical architecture has proven disastrous, producing ugliness, crowding, pollution and dehumanization in the cities, which have then spawned suburbs, with the entire system causing ecological imbalance on the rest of the planet.

Tall buildings have always been a manifestation of pride, cruelty, power, or ponderous religiosity. The Tower of Babel, of course, is the archetype. But we also have the sacrificial altars of the Aztecs, the pyramids, the cathedrals, and the bastions of big business today.

To satisfy this phallic impulse, we might build an occasional skyscraper, perhaps even a mile-high tower such as Wright designed, just to show that we can do it, or to serve as landmarks, or as places from which to have a splendid view. Otherwise, architecture should maintain a low profile.

There are some signs that people are beginning to understand this, but unfortunately it has not yet emerged as a popular issue. All that has appeared so far is an occasional protest against a gratuitous tower, or something like the petition leading to Levi's Plaza in San Francisco, where the company's employees demanded, and got, a "low rise" complex, thus preserving some of the last views still left in the heavily overbuilt downtown area.

However, short of implementing the ideas in Ayn Rand's *Fountainhead*, or a succession of earthquakes and tidal waves, or nuclear war, the monstrosities of vertical architecture will be with us for a long time.

Lying down is not often thought of in terms of human rights and political freedom, for it would seem that no matter what other form of oppression might be imposed, a person could always at least lie down.

But this isn't always the case. Denying someone space in which to lie down has been and is being used as an insidious form of punishment. The notorious tiger cages in South Vietnam were a prime example of torture through removing a prisoner's liberty to stretch out. And in *Prisoner Without a Name, Cell Without a Number*, Jacobo Timerman describes, with a dispassion that makes his ordeal all the more sobering, just such a condition.

"The cell is narrow. When I stand at its center, facing the steel door, I can't extend my arms. But it is long, and when I lie down, I can stretch out my entire body. A stroke of luck, for in the cell I previously occupied—for how long?—I was forced to huddle up when seated and keep my knees bent while lying down.

"The floor of the cell is permanently wet. Somewhere there's a leak. The mattress is also wet. I have a blanket and to prevent that from getting wet I keep it on my shoulders constantly. If I lie down with the blanket on top of me, the part of my body touching the mattress gets soaked. I discover that it's best to roll up the mattress so that one part of it doesn't touch the ground. In time, the top part dries. This means, though, that I can't lie down, but must sleep seated. My life goes on during this period—for how long?—either standing or seated."

The imposition of verticality as a physical punishment, however, pales before the psychosocial regimentation of prison systems as a whole since the appearance of the Panopticon, the vertical tower from which all prisoners could be monitored down to their most private behavior. Foucault's description of Bentham's invention is chilling, but not nearly so much as the implications of the idea, which was but an institutional application of the more general process of substituting the human eye of punishment for the divine eye of judgment.

In the eighteenth century, with the dawning Age of Reason making its influence felt and the impending Death of God in the wings, it was necessary to find a new means of restraint for keeping individuals obedient to the laws of society. The Panopticon was the first crude effort that led ultimately to the entire contemporary world of surveillance, with its ubiquitous television cameras, wire taps, and satellites with powerful lenses peering down from the skies.

This vertical heightening found its most sinister tool in the development of the superego, with which each person censors himself or herself continually, so that we are all turned into self-contained prisons.

Lying down as art. The vertical understanding of art is as creation: making things. The horizontal understanding is one of appreciation: becoming more sensitive to what is. The slogan "art is dead" is true enough of the creative approach, which can do little more than put old wine in new bottles, e.g., "computer art." But that whole movement can be summed up with a paraphrase of the complaint in Ecclesiastes some 2500 years ago, "Of making many works of art there is no end."

There are two aspects to the horizontal understanding of art. The first involves the probability that the origins of art lay in an awareness of the body as object. This awakening occurred during our "magical" period, when the species was beginning to differentiate itself from nature. Rank quotes E. van Sydow to the effect that "The beginning of art lay in its application to the body. Art rose, not in any isolated and self-contained work, but by the moulding of the human body to a formative plasticity, urged thereto by an idealized instinct of will to style."

Art may very well have begun with tattooing, scarification, body painting, the making of masks, and dance and song, all growing out of the awareness of the body as body, distinct from other bodies or nature at large. Dessoir notes that "The actor nowadays represents this pristine type of artist, where object and subject coincide, and the body forms the material in which and through which the artist creates."

In this model, art followed a progression from an awareness of the body, to using the body as material, and finally to creation of objects by the body—an unmistakable vertical process. From the viewpoint of the horizontal paradigm, then, art would be returned, down the ladder of creation, back to simple awareness, first of such functions as breathing, posture, and the senses, and then of all the other qualities and textures of the organism and of the world altogether. In this context it would not be necessary to make anything new, but only to enjoy in depth what already is.

This horizontal shift has already occurred in the avant-garde of art sensibility, beginning with Duchamp who signalled the change from art as creation to art as appreciation, and finding its complete statement in the approach of post-modernist artists, as reflected in Lucy Lippard's *Six Years: The Dematerialization of the Art Object.*

"Sol le Witt's premise that the concept or idea was more important than the visual results of the system that generated the object...was soon applied to such ephemeral materials as time itself, space, nonvisual systems, situations, unrecorded experience, unspoken ideas, and so on. Such an approach to physical materials led directly to a similar treatment of perception, behavior, and thought processes per se. The most effective method in this case has often been the accent or overlay of an art context, an art framework, or simply an art awareness...on existing situations or information."

Along the same lines, she quotes Robert Smithson: "As long as art is thought of as creation, it will be the same old story. Here we go again, creating objects, creating systems, building a better tomorrow."

As a pure exercise in art as appreciation, Lippard includes a "piece" by Rick Barthelme:

> Being—in the physical condition—facing north;
> Being—in the physical condition—facing south;
> Being—in the physical condition—facing east;
> Being—in the physical condition—facing west.

Thus, on considering the vertical to horizontal shift in art, one finds a re-awakening of the fact of the body not only as the matrix but as the very substance of art, and toward appreciation of what is, instead of the creation of more artifacts.

This doesn't mean we can't keep making paintings and pots and songs, but that such creation be born of spontaneous expression or an esthetic pragmatism, an approach summed up both by the Balinese with their saying, "We have no art; we try to do everything the best we can," and the Tibetans who state, "The highest art is the art of living an ordinary life in an extraordinary manner."

We've had stand-up comedians, why not lie-down comedians? The vertical comic puts himself in a position of vulnerability by confronting the audience and having to "deliver." This insecurity produces an aggressive defensiveness or a cutting edge, exemplified by the model of the "punch" line and the real and symbolic pratfall. What sort of humor would a horizontal comic have? What sort of stories and jokes does one tell while lying down?

What if the Johnny Carson show were done with everyone reclining? Why not have the studio audience lie down also? After all, aren't most of the people watching at home lying in bed or on a couch? If the goal of the comic is to "lay them in the aisles," why not just invite people to do that instead of first cranking everyone up through expectation? Vertical humor at its best is cathartic. Horizontal humor would be nourishing, tending more toward soft smiles than sharp laughter.

Along with sleeping, sex is the activity most commonly associated with lying down. Murray S. Davis describes the connection neatly in *Smut: Erotic Reality/Obscene Ideology.*

"The private arena where erotic reality is most often generated is the 'bedroom.' 'To go to bed with' is another of the most frequently used euphemisms for sexual intercourse. Those who go to bed together for sex shift from vertical to horizontal interaction, a form of interaction that occurs in very few other activities (sunbathing being perhaps the most common). Many rules governing the behavior of the 'upright' are suspended for the prostrate, whether they be relaxing, fucking, giving birth, drunk, sleeping, sick, dying, or in the grave. The horizontal position itself implies sexual activity. The terms 'lie with,' 'lay,' 'tumble,' 'upend,' and 'go down,' all have sexual connotations. In Shakespeare's *Henry IV*, Hotspur describes Kate as 'perfect at lying down.' "

Actually, a good deal of the difficulties that many couples have sexually comes from their not making a distinction between sleeping together and having sex on the one hand, and lying down as an act in its own right on the other. The first two activities are either personal or impersonal, while the latter is transpersonal. And unless two people who are coupled establish a strong transpersonal bond, their personal and impersonal contacts can very easily become ingrown or alienated.

It would be helpful for couples to experiment with lying down as an activity independent of sexing and sleeping, letting it be the matrix within which both making love, and sharing dreams and oblivion, arise. In addition, each individual might lie down alone, either at the same time or a different time than the other, in the first instance as a spiritual communion and in the second instance as a conscious separation, a return to pristine solitude.

The relationship between sex and the horizontal is obvious on the physical plane, but in the psycho-emotional realm approaching sex horizontally is very difficult in our culture. As one person put it, "How can I be laid back and still get laid?" There is no easy answer to that question since it addresses complex issues of existential anxiety, the lifestyle of civilization, and the making of distinctions between sex, metasex and eroticism. However, the question can be approached and explored by using lying down as a generator, amplifier and modulator of desire, prior to foreplay, as foreplay, as bass line during the rhythmic movements, and in afterglow.

With lying down it is possible to experience some of the effects associated with various "mind-altering" substances. If you plan to combine any of these with lying down, however, it's a good idea to take into account the fact that each has its own directional quality.

LSD, for example, is a vertical substance. It takes you up and down the ladder of hierarchical consciousness. Ideally, it should push you out the "top" of your mind, but very often it results in an unsettling tour of the "lower" levels. When one lies down on LSD, the experience becomes more somatic and non-conceptual, allowing for the release of feeling states.

Marijuana is a horizontal substance. It provides a unique skew in perception and knocks the ducks of linear thought out of their row. Lying down after smoking or eating grass reinforces the horizontality of the marijuana effect and one tends to drift, dissociate and fantasize freely.

Peyote, psilocybin mushrooms and mescaline are diagonal substances. They operate like light filters with a diagonal bias which, when placed between a horizontal and vertical filter, allow otherwise unpassable light to shine through. Lying down after ingesting these substances produces "rushes" in the organism which are like winds blowing open the doors of perception.

Alcohol is a completely disorienting substance, wiping out both vertical and horizontal reference points. Lying down while intoxicated produces dizziness, spinning, a sense of toppling backward into the void. As Dean Martin noted, "If you can lie on the ground without having to hang on, you're not really drunk."

G.K. Chesterton is reputed to have said, "Whenever I get the urge to exercise, I lie down until it passes." He might have been pleased by an innovation in exercise made possible by a technology not available in his day, through which one can lie down and be exercised at the same time. This is a system, now offered by a number of spas, of toning the muscles through electrical stimulation. Electrodes are attached to various parts of the body and as a current is passed through them, the underlying muscles contract. One ad runs, "Lie back and enjoy your fantasies while our computers tone your muscles to the equivalent of 1500 situps, 1500 pushups, and 1500 chinups."

This exercise for the lazy is but one manifestation of the exercise craze that has seized the psyches of tens of millions of people in the

United States, and which constitutes a secular fundamentalist religion for followers of calisthenic charismatics like Richard Simmons and Jane Fonda. But being electrically massaged by a computer while you are lying down is as much a parody of real exercise as what one finds on morning television or in books by shapely celebrities.

For people who work at sedentary occupations, "exercise" is obviously necessary. But since it is intrinsically a vertical activity, there is little to be said about it from the horizontal viewpoint, except to suggest that any exercise period begin with several minutes of lying down so that the ensuing movements are more grounded and graceful.

Paradoxically, the most horizontal approach to exercise has come through the introduction of a device which produces a vertical posture. This is the gravity boot, which allows one to hang upside down from hooks attached to the ankles. While suspended in this way, one's posture is perfectly vertical—a plumb line to the center of the Earth—but since one is surrendering entirely to gravity with complete relaxation, the whole sensibility and effect is that of horizontality. Actually, the suspension might be called a para-horizontal posture.

In his book *The Gravity Guiding System*, Dr. Robert Martin points out that the average person spends two-thirds of his or her life in the "dominant" postures of sitting or standing. As a result, over time, the weight of the upper body imposes a force that compresses the spine from end to end. He recommends the exercise of hanging upside down to counteract this and all the other ills we've bought into as a result of becoming vertical mammals.

About the horizontal posture he writes, "We begin our existence in the horizontal position, and every day of our life from then on we resort to this position for rest and relaxation. Little effort on the part of our muscles is required in this position. Few or no demands are made on the nervous system and the musculo-skeletal structure in maintaining equilibrium. Horizontal is the most neutral of all postures. Circulatory resistance is minimized; therefore, vascular prolapsis is relieved."

In reviewing three paintings of reclining women, two by Sargent and one by John White Alexander, Linda Nochlin describes the worldview represented by those works:

"Repose! The very word has a nostalgic ring to it, conjuring up a vanished world of pale solitude, gentle distances, summer vistas, languour, and lovely women... Our current ideals of beauty and desireableness are so intensly dynamic, equated with jumping, running, kicking up the heels, stretching and bending in the photographs of fashion journals, our notions of success so stringently identified with frenetic action and assertiveness, that it is hard to recapture the spiritual outlines of a world in which repose, instead of being a kind of blasphemy...was, on the contrary, considered a basic and charming feminine virtue."

Nochlin rightly points out that in the historical period she is describing, languour definitely had overtones of decadence in the class of women in whom it was considered a virtue. "The notion of repose clings insistently to the *fin de siecle*, when *luxe* and *calme*, if not *volupté*, were considered plausible elements of the good life—for the privileged few who could afford it." She doesn't mention it, but the fact that men are left out of the equation altogether implies an even greater bias.

The question all of this raises is why something so profoundly enjoyable should be relegated to the status of a luxury for the upper class, and the women of that class in particular. Why shouldn't being at ease, even voluptuously so, be the ground of social life instead of a preserve for the wealthy, and even at that only in certain times? Wouldn't civilization be relieved of its discontents if it inclined toward the horizontal, which Nochlin calls "this most soothing of all directions"?

In his book *Entropy*, Jeremy Rifkin offers up the second law of thermodynamics as a contender in the new paradigm sweepstakes. The Entropy law, as is well known, states that matter and energy can only be changed in one direction, that is, from usable to unusable, from available to unavailable, or from ordered to disordered. In local terms, such as apply to the Earth and the various ecosystems on it, entropy is "a measure of the extent to which available energy in any subsystem of the universe is transformed into an unavailable form." Burning wood for heat and being left with ashes is an obvious example.

When population is small, technology simple, and renewable energy sources, such as wood, are used, entropy is kept to a minimum. In our time, however, population is enormous, technology massive, and our major sources of energy are non-renewable. All of this has led to a global entropy watershed, a condition aggravated by humanity's still being geared to "maximum energy flow-through."

This last concept refers to the fact that each species and every organism functions as an energy transformer. All creatures stay alive by the process of energy flow-through—food in, waste out. In the early stages of an ecological system's development, energy flow-through is maximized, since there is still an excess of available energy present. However, "as various species begin to fill up a given ecological habitat, they are forced to adapt to the ultimate carrying capacity of the environment by using energy flow-through more efficiently. The early stage of maximum flow-through is generally referred to as the colonizing phase, and the later stage of minimum flow-through as the climactic phase."

Rifkin continues,"On the whole, Homo sapiens has yet to move from a colonizing to a climactic phase. Human beings, especially in the highly developed industrial societies, continue to order activity in such a way as to increase energy flow-through in both the human and social systems. The worldwide human crisis today is a crisis of transition. In the next age, humanity will have settled into its climactic phase, ordering its activity in such a way as to minimize energy flow-through in the human and social processes. If it doesn't, it will likely go the way of other species who were unable to make the transition in the past. Life's epic is strewn with extinct species; it would have little trouble adding one more to the long list of names."

In the past, when a people arrived at an entropy watershed, they either migrated, found new sources of food and energy, or introduced new technology. Now, however, the problem is planetary. As Rifkin notes, "The human population is doubling every forty years, and every nook and cranny on the entire globe appears to be filling up with standing room only. We find it harder and harder to locate sources of available energy and more and more difficult to find places to discard our energy wastes. We are finally reaching the outer limits of the planet Earth, and as we begin to jostle each other back and forth, a new voice can be heard from deep inside the crowd. The voice is getting louder and it's saying that we must learn to *live within our limits*. The colonizing stage for Homo sapiens is truly over and the transition from a colonizing to a climactic mode of existence is the most profound change our species will ever have to make. The crossroad is now before us."

What is necessary is that we do globally what some cultures have done locally. "Throughout history," Rifkin points out, "there have been exceptional cultures which have been able to survive for long periods of time within the same energy environment. They were able to make the transition from a colonizing to a climactic stage of existence. For all of these cultures, adapting to the existing environment meant slowing down the flow-through of energy and thus slowing down the entropy increase in the overall environment. For them, 'living within their limits' was a matter of second nature."

Much of the book is a detailed accounting of the ways in which planetary civilization is, in fact, moving away from this kind of sane perspective, and has instead become overgrown, top-heavy, dependent on an accelerating plunder of the world's resources, and produces waste and pollution that is approaching lethal levels. The United States is singled out as the worst offender, with its "addiction to energy" manifesting in a notorious statistic—with 6 percent of the world's population it consumes over 33 percent of the world's resources.

Rifkin ends the work with an exhortation. "The alternative to this wholesale squandering of available energy is an internalization of the values and dictates of the entropic paradigm... The first step in this historic process is to fully comprehend what it is, as people, we believe... We must also join together, in a popular, grassroots social force, to begin the dismantling of the existing high-energy infrastructure. At the same time, we must build our new society based on a new set of values which reflect our awareness of the entropy process."

This political program is tied, at last, to a spiritual principle. "Preserving and enhancing life in all its forms requires available energy. The more energy available, the greater the prospects for extending the possibilities of life into the future. But the second law also tells us that the available store of energy in the world is continually being depleted by every occurrence. The more energy each of us uses up, the less is available for all life that comes after us. The ultimate moral imperative, then, is to waste as little energy as possible. By so doing, we are expressing our love of life and our loving commitment to the continued unfolding of life."

All of this is splendid vision, analysis and rhetoric, and I would take only one exception to it—equating the old paradigm with the Newtonian worldview. That's too limited. The "machine universe" model is only a few hundred years old and is itself but a final flourish of a more profound and radical paradigm, that of verticality, which is the essential species worldview out of which all local vertical cultures and concepts emerge.

Otherwise, Rifkin is right on the money. The transition he describes from a colonizing to a climactic phase is clearly the shift from a vertical to a horizontal mode of managing the economics and ecology of energy. And the only possible response to the implications of entropy is lying down.

Psychologists, who have taken human behavior, feeling and thought as their domain, have almost entirely overlooked the importance of posture in determining what it means to be human in the first place. They have studied everything from the molecular structure of neurons to mob frenzy, but seem to have been completely blind to the fact that everything from the size and complexity of the brain to the grossest forms of activity are functions of our having stood erect.

One brilliant exception to this collective disregard for basics appears in the essay, "The Upright Posture," by Erwin Straus. The piece offers a rare appreciation by a Western scientist of the upright posture in the shaping of human consciousness. It is so rich in detail and suggestion that making any selection from it is difficult, but the following will give at least some idea of Straus's vision.

He begins by noting that the "upright posture is...pregnant with a meaning not exhausted by the physiological tasks of meeting the forces of gravity and maintaining equilibrium. The upright posture distinguishes the human genus from other living creatures. It does not occur in any species other than man. Upright posture pre-establishes a definite attitude toward the world; it is a specific mode of being-in-the-world."

This notion might be labeled physiological existentialism, for it links the most profound considerations on the human condition directly to the vertical posture, beginning with the very notion of awakeness itself. "When we stand erect," he notes, "Gravity is never fully overcome; upright posture always maintains its character of counteraction. It calls for our activity and attention. Automatic regulation alone does not suffice. An old horse may go to sleep standing on its four legs; man has to be awake to keep himself upright. Much as we are part of nature with every breath, with every bite...we first become our true selves in waking opposition to nature. In sleep, we do not withdraw our interest from the world so much as we surrender ourselves completely to it. We abandon ourselves to the world, relinquishing individuality. We no longer hold our own in the world, opposed to it. Awakeness and the force of gravity are mutually interdependent."

In sum, the moment we became upright creatures, then the bare act of standing—which was now our "mode of being-in-the-world"—required an effort of wakefulness. But this wakefulness was subject to the very force it was meant to overcome. As Straus puts it, "While awakeness is necessary for the upright posture, that is, for counteracting gravity, gravity determines waking experience. The dreams of one night are not related to the dreams of another night, but days are related to one another. They form a continuum where every hour, every moment, anticipates the next and prepares for it. Held back by gravity to a precise point, we can overcome distance only in an orderly sequence. During our waking hours, sequence means consequence. Gravity, which holds us in line, imposes on waking experience a methodical proceeding."

Thus, the human, by standing erect, commits himself and herself to a perpetual state of struggle or negotiation with gravity, which is the state of awakeness. But this awakeness is then shaped and structured by gravity, transforming it into consciousness, which is immediately beset by insecurity. Upright posture is something "we learn in and through falling," Straus points out, and it "remains threatened by falls throughout our lives. The natural stance of man, therefore, is resistance. A rock reposes in its own weight. The things that surround us appear solid and safe in their quiet resting on the ground, but man's status demands endeavor. It is essentially restless. We are commited to an ever renewed exertion."

Given this resistance and unrest as the primary psychological manifestations to emerge from the existential consciousness resulting from the upright posture, the sociological implications are clear. "Considering man in his upright posture, we do well to envisage the possibility," Straus continues, with proper academic reserve, "That it is not society that has first brought man into conflict with nature, but that man's natural opposition to nature enables him to produce society, history and conventions. The direction upward, against gravity, inscribes into space world regions to which we attach values, such as those expressed by high and low, rise and decline, climbing and falling, superior and inferior, elevated and downcast, looking up to and despising."

In another section of the essay, Straus examines another result of the upright posture, ambivalence. "With upright posture," he writes, "An inescapable ambivalence penetrates and pervades all human behavior. Upright posture removes us from the ground, keeps us away from things, and holds us aloof from our fellow men. All of these three distances can be experienced either as a gain or a loss."

In the first aspect, distance from the ground, we gain "freedom of motion and enjoy it, but, at the same time, we lose secure contact with the supporting ground, with Mother Earth, and we miss it. We stand alone and have to rely on our own strength and capacities." Straus further notes that children in their early years use their given name to refer to themselves, but that when they reach the age of being able to stand firmly on their own two feet, they begin to use the pronoun "I."

In the second aspect, distance from things, Straus points out that "a child creeping on his hands and knees not only keeps contact with the ground but is, in his all-fours locomotion, like the quadrupeds, directed toward immediate contact with things. The length-axis of his body concides with the direction of his motion. With getting up, all this changes. In walking, man moves his body in a parallel transposition, the length-axis of his body at a right angle to the direction of his motion. He always finds himself 'confronted' with things. Such remoteness enables him to see things, detached from the immediate contact of grasping and incorporation, in their relation to one another.

Seeing is transformed into 'looking at.' The horizon is widened, removed; the distant becomes momentous, of great import. In the same measure, contact with near things is lost."

And in the third aspect, distance from fellow men, the vertical takes its toll as well. "In upright posture, we find ourselves 'face to face' with others, distant, aloof—verticals that never meet. On the horizontal plane, parallel lines converge toward a vanishing point. Theoretically, the vanishing point of parallel verticals—to which we are comparable, standing vis-a-vis—is in infinite distance. In the finiteness of seeing, however, parallel verticals do not meet. Therefore, the strict upright posture expresses austerity, inaccessiblity, decisiveness, domination, majesty, mercilessness, or unapproachable remoteness, as in catatonic symmetry."

In conjunction with the existential, psychological, sociological and attitudinal transformations brought about by the upright posture, there are the anatomical developments and their implications for the definition of human nature. The three major changes are the formation of the head with its large-brain capacity, the freeing of the arms and hands, and the reorientation of the oral-anal axis.

"Upright posture has lifted eye and ear from the ground," Straus says in talking about the formation of the human head. This means that seeing and hearing assume dominion over smell, which is the first-born of the senses, and so our attention is turned from texture to distance, or from the concrete to the abstract. Also, "With the upright posture, with the development of the arm, the mouth is no longer needed for catching and carrying or for attacking and defending. It sinks down from the 'visor line' of the eyes, which can now be turned directly in a piercing, open look toward distant things and rest fully upon them, viewing them with the detached interest of wondering. Bite has become subordinate to sight." Among other things, these changes, coupled with the increased size and complexity of the brain, allowed the development of speech and thought. Or, as Straus puts it, "sight becomes insight."

Concurrently, the arms and hands undergo transformation. "In upright posture, the front extremities are no longer asked to support and carry the body. Relieved from former duties, they're free for new tasks. The anterior limbs develop into the human arms and hands and acquire multifarious new functions." Straus describes the obvious advantage of the flexibly opposable thumb but states that even more important is the hand's becoming "an organ of gnostic touching—the epicritic, discriminative instrument par excellence. As such, the hand now ranks with the eye and the ear." And in discussing the relationship between the hand as sensor and the perception of "experienced distance," Straus weds his physiological existentialism to a zen-like vision. "Distance is ambivalent. Sometimes we want to preserve it, sometimes we want to eliminate it. The hand is instrumental in both cases. When our equilibrium is out of balance or disturbed, the hand grasps for a hold. In darkness, it functions as scout and sentry, warning against collision and searching for contact. Sometimes no hold is found, no contact is made. Searching hands stretch into the void. It is as if emptiness were localized in our hands. Indeed, only the empty hand, like the beggar's hand, can receive. Emptiness is the condition by which our hands can be filled. Only because of remoteness can they make contact."

The third major development resulting from the upright posture is our removal from the functional aspects of sex and food. Straus doesn't discuss the sexual changes, but they can be pretty easily discerned. In standing erect we had to open the pelvic girdle outward, thus producing fleshy buttocks behind and allowing face-to-face intercourse. This introduced the secondary elements of sensual and personal stimulation into the straightforward business of procreation, which led to the virtual elimination of the oestral cycle as the determining factor in frequency of sex, and to the weakening of purely bio-genetic impulses as the determining factor in the choice of mate or mates. All of this made sex arbitrary and confusing, thus requiring regulation via cultural taboos and societal restrictions. At the same time, our orientation toward food and feces mutated. "Animals move in the direction of their digestive axis," Straus notes. "Their bodies are expanded between mouth and anus as between an entrance and an exit, a beginning and an ending. The spatial orientation of the human body is different throughout. The mouth is still an inlet but no longer a beginning; the anus, an outlet, but no longer the tail end. Man in upright posture, his feet on the ground and his head uplifted, does not move in the line of his digestive axis; he moves in the direction of his vision. He is surrounded by a world panorama, by a space divided into world regions joined together in the totality of the universe. Around him, the horizons retreat in an ever growing radius. Galaxy and diluvium, the infinite and the eternal, enter into the orbit of human interests."

This definition of humanity as a function of the upright posture is brought together with a concluding paragraph. "The sensory organs cannot change without a corresponding change in the central nervous system. No part could be altered alone. With upright posture, there is a transformation of sensorium and motorium, of periphery and center, of form and function. While upright posture permits the formation of the human skull and, thereby, of the human brain, the maintenance of upright posture demands the development of the human nervous system. Who can say what comes first and what comes last—what is cause and what is effect? All these alterations are related to upright posture as their basic theme."

Because of the length of Straus' essay, I've been able only to quote a few passages, but I think these are sufficient to give strong support to the thesis that since our inception as a species we have been subject to the single paradigm of verticality. For Straus, however, the matter ends there. He obviously sees verticality not only as our origin but as our destiny, for better or for worse, with no possibility of future evolution except in the same direction.

This is clear in his closing comments on horizontality. "Because getting up and standing are so demanding, we enjoy resting, relaxing, yielding, lying down, and sinking back. There is the voluptuous gratification of succumbing." But lest we be lulled by these seductive images into a horizontal attitude, he pulls us up short with the following admonition. "The phenomenon of upright posture should not be neglected in favor of the lying man, or the man on the couch. To do this is to ignore facts that are obvious and undebatable, accessible without labyrinthine detours of interpretation, facts which exact consideration and permit proof and demonstration. It is true that sleep and rest, lying down and lying with someone, are essential functions; it is no less true that man is built for upright posture and gait, and that upright posture, which is as original as any drive, determines his mode of being-in-the-world."

From his tone, there is, to me at any rate, the sense that Straus saw the edges of the horizontal vision, and then drew back. To have made the shift, all that would have been necessary is changing the tense of the final verb, from "determines" to "determined." This doesn't mean accepting that we can change our anatomy, but understanding that we can change the metaphors which our anatomy has bequeathed us.

The way to do this is through lying down in its four fields, of which the metaphoric is the third. Here we assume an as-if approach, and act as though we could withdraw our essence, as it were, from the vertical principle, and inhabit our upright posture in purely functional terms. The vertical stance would then become simply a tool for getting the necessary survival tasks done, but it would no longer dominate our "being-in-the-world." We would be defining ourselves existentially as horizontal beings even while, phenomenally, we went on using our eyes, hands and genitals in an upright manner. Our evolution would continue as a subtle transmogrification of intelligence within an increasingly translucent and elegant vertical frame, until our understanding outshone our finally harmonious functioning.

Freud's greatest contribution was horizontal, but it was never accorded its proper weight, either by himself or anyone who came after. This was his asking the patient to lie down, an innovation that was given second-class status as one of the aspects of technique, which itself has been considered not nearly as important as theory. This vertical conceptual bias obscured the significance of the posture, keeping analysts glued to the "material" that is produced when a person lies down instead of focussing them on the actual transformations brought about by the posture.

In *Three Worlds of Therapy*, Anthony Barton describes some of the functions of lying down in the process of psychoanalysis. "That the patient lies down provides a number of advantages. First, the patient is relaxed and more susceptible to saying whatever comes into his mind because the position itself invites a loosening of the rational structures of a hyperrational consciousness. Lying down also has the effect of making many of the usual social amenities seem irrelevant. It immediately undercuts many specific social conventionalities for the patient, for it is difficult to be pompous, to make skillful party chatter, or to lecture while lying on a couch...

"Because lying down is a position for sleeping, it also encourages the emergence of the less conscious, more fantastic side of the personality. The patient is relaxed, afloat, and adrift, and to a certain extent cut off from reality. He is literally not facing the world and is therefore not reality oriented in the everyday sense of the term.

"Finally, the patient's lying down partially erases the distinction of high and low, the differentiation of the world into above and below. He allows both high and low to speak equally to him. Lying down, then, invites relaxation, a loosening of control, a diminishing of reality coping, and a certain unconventionality of social expectation...

"The analyst is also affected by the patient's lying down. That the patient lies down and the analyst does not is one of the most important aspects of the analyst-patient relation, because the analyst is then confirmed as being in a position of superiority and priority as regards the interpretation of reality. He is one who has his feet on the ground and can see far... The analyst is the 'seer' and the patient 'laid out before him' is the relatively passive producer of data to be understood and seen. The analyst is also affected by an empathic possiblity for regression in himself. The fact that the patient lets down his controls helps the analyst to loosen his own control empathically— to be easy and free-floating with the patient because the patient is easy and free-floating with him."

This is a succinct and accurate description of the value of lying down as a facet of analysis, but it maintains the conventional perspective on the posture as only a tool for the more important process of verbal therapy, whereas I am suggesting that lying down is itself the "therapy."

The vertical bias of analysis is clearly shown in the analyst's remaining seated in order to be in a "position of superiority and priority as regards the interpretation of reality." But why should it not be possible to understand reality as well, if not better, from the horizontal? That no one has ever asked that question in analysis, much less considered that the truths of horizontality are not less than the truths of verticality, shows the depth of the prejudice, the glaring blind spot.

Also, all of the virtues of lying down that Barton so insightfully enumerates are treated by analysts as temporary turnouts on the highway of interpretation. "Relaxing," allowing "both high and low to speak equally," and finding it "difficult to be pompous," are viewed by analysts as mere mediums for the flow of the self-referential monologue, instead of as the very condition of cure that the analysis is aiming at many years and tens of thousands of dollars down the road.

It's fascinating to speculate how the development of analysis, of psychology, and indeed of the twentieth century would have been different if Freud had decided to lie down with his patients (not on the same couch, of course!). One grows almost giddy imagining a Freud who could stretch out, relax, breathe deeply, and chat amiably and brilliantly with his patients, a Freud who really became "easy and free-floating." At the end of the session he might say to the patient, "You're beginning to get the feel of being at home with yourself. Carry that with you. That's the real healing."

And not a *word* about what mummy and daddy were doing the night you stood at the bedroom door...

Physicists have a way of seeing themselves at the cutting edge of knowledge, the current avant-garde of the vertical thrust of humanity into the mystery of the universe. They attempt to go as far back into time, as far out into space, and as far deep into energy and matter as it is possible to go.

But they are always being humbled by their own discoveries and foiled by their own methods, which gives the best of them a sense of humor and irony, and makes the denser ones all the more determined to find the "final equation."

In 1900, Lord Kelvin noted that there were only "two clouds on the horizon of physics." These were the problem of black-body radiation and the result of the Michelson-Morley experiment. He informed the Royal Institution at the time that there was no doubt that the troublesome discrepancies would soon be explained away. He was right; they were. But in a way that caused the Olde Paradigm of Newtonian physics to be entirely eclipsed. Planck uncovered quanta and Einstein introduced relativity.

The new wave of physics was wildly successful for a few decades and its adherants, like Lord Kelvin during the previous era, were convinced that it wouldn't be long before they understood everything. Again, however, there appeared a few clouds on the horizon, the most notorious of which was named after the man who spotted it, Heinsenberg. Gary Zukav nicely describes the nature of the problem it presented. "Heinsenberg's remarkable discovery was that there are limits beyond which we cannot measure, accurately, at the same time, different processes of nature. These limits are not imposed by the clumsy nature of our measuring devices, or the extremely small size of the entities that we attempt to measure, but rather by the very way that nature presents itself to us. In other words, there exists an ambiguity barrier beyond which we never can pass without venturing into the realm of uncertainty."

In addition to this thunderhead, other clouds include an embarrassing proliferation of "elementary" particles, the fact that the mathematical foundation of contemporary physics—randomness—cannot be mathematically defined, and the sci-fi like possibility that there are "worm holes" in the cosmos leading to other universes.

Physicists are only human, however, and hope springs as eternal in their breasts as in anyone's. Also, they generally view mystery the way many doctors look at death, as a minor, and perhaps even solvable, difficulty in an otherwise splendid theoretical framework. So they continue to seek the ultimate particle, which is also a wave, which turns out to be an abstract relationship, and they continue to try to formulate the grand synthesis which will once and for all unify everything. And all of it would be the stuff of a Swiftian satire except for the terrible weapons that are made from the symbols they scribble on their pads.

One wonders at what a horizontal physics would be like. It would not necessarily use different methods, but it would have a more sophisticated understanding of what it was up to. It would accept that knowledge is intrinsically limited, curved in on itself, and approach its quest more as a kind of pastime than a straight-faced forging fiercely ahead to capture creation in a net of ideas.

Prostitutes were called "the Great Horizontals." And Huey Newton, when asked the place of women in the revolution, replied, "On their backs."

Such references make the discussion of the relationship between women and lying down potentially a bit touchy. But there is a critical link here. For example, the overwhelming majority of teachers who use lying down as a method or technique are women. Certainly the greatest teachers of this healing body-art are women, e.g., Charlotte Selver, Laura Huxley, the late Else Gindler.

This "school" is almost entirely unknown outside the circles of its adherants, however, and for several reasons. The first is that these women don't sell their wares like their male counterparts, the gurus, yogis and "holy men." They don't form organizations of worshipful followers, nor crank out an endless stream of books filled with photographs of themselves.

The second reason is that teaching awareness and intimacy in this society is a chancy business. Our civilization is drastically vertical and hard-edged. A person relating and responding in this world using a horizontal approach may very easily be victimized, institutionalized, or simply not be seen at all. So students of the horizontal tend to be cautious and reserved, the opposite of the prevailing public relations and advertising mentality that dominates the culture.

Finally, few of the teachers in this field realize the full implications of their work. Many of them have been conditioned by the hierarchical prejudice of Western thought in which awareness is considered a low-level function of sentience, as opposed to "consciousness," which supposedly makes us superior to the other creatures on the planet. So these women generally view their classes as little more than adjunctive therapy, for reducing stress and producing mild insight. As such, they perform a humanitarian service in helping people relax, but they miss the radical aspect of lying down, its power as a paradigm.

In the area where women have claimed a leading revolutionary role, on the other hand, in the long war known as Women's Liberation, women have used vertically aggressive methods almost exclusively. The result is that, for whatever gains they have made, they have paid with schisms and splits often marked by bitter infighting. Also, many of its prominent heroines have acquired the very stridency, narrowness and rigidity for which they attack men.

What the alternative might be, in practical terms, is unclear. The patriarchy is on its last legs, but there can't be a return to the matriarchy either—that phase of evolution is also complete. Ideally, we need to enter a holoarchical stage of understanding in which both genders, while retaining their biological particularity, move toward the overall emergence of an androgynous mind.

The *Tao Te Ching* says that "the female overcomes the male with stillness, lying low in stillness," which suggests a futurist fantasy about women and lying down. It is that a global movement takes place in which all the women on the planet are joined in an empathic bond, and simultaneously resign en masse from all vertical enterprises. In the sudden ensuing Gurdjieffian "shock," the men might notice the complete absurdity of their vertical games, and perhaps take the first deep breath they've collectively enjoyed in more than ten thousand years, and then decide to lie down also.

Then, lying down, women and men might let all the agitations and tribulations of vertical civilization settle like silt on the bottom of a pond, and in the quiet water see the reflection of a new quality of being. In that awareness we could all stand up together again and see what real work was at hand.

Annie Dillard writes: "Scholarship has long distinguished between two strains of thought which proceed in the West from human knowledge of God. In one, the ascetic's metaphysic, the world is far from God. Emanating from God, and linked to him by Christ, the world is yet infinitely other than God, furled away from him like the end of a long banner falling. This notion makes, to my mind, a vertical line of the world, a great chain of burning. The more accessible and universal view, held by Eckhart and many peoples in various forms, is scarcely different from pantheism: that the world is immanation, that God is in the thing, and eternally present here, if nowhere else. By these lights the world is flattened on a horizontal plane, singular, all here, crammed with heaven, and alone..."

The argument between those who uphold a vertical God and those who accept a horizontal God has continued for thousands of years, and even produced a third school, generally known as the

Perennial Philosophy, which holds that God is beyond all categories, including immanence and transcendence, and can only be discussed in terms of paradox.

The philosophic merits of these various viewpoints aside, each of these approaches has produced a distinctive way of life Those who incline toward paradox are mostly intellectual and reclusive, monkish. Those who embrace a horizontal understanding are usually "native" peoples. While those who worship a vertical God are at best smug and at worst intolerant and bellicose.

Christianity provides the best example of this last case. During its first fifteen hundred years it systematically stamped out all nature religions under its rule, incorporated and subsumed all the local religions of towns and villages, and declared paradox to be a heresy, imposing its hierarchical dogmas through enforced acculturation, intimidation and torture. Then, when it burst loose from Europe, it travelled at the head of rapacious armies in order to subjugate and convert the "heathen."

How such a bigoted and bloodthirsty religion can have arisen from the life of someone like Jesus is a supreme example of irony in history. There are several reasons for it. One is that the religion wasn't founded by Jesus but by the thug Paul. The second is that Jesus was killed while he was still young and in the fiery stage of his mission. If he'd lived as long as Gautama and had time to grow mellow, we would have been left quite a different gospel than the one we have now. The cursing of the fig tree, the endless esoteric exhortations to "sleep not," the railings against the bureaucratic Pharisees, and all the other explosions of youthful energy would have been balanced in later years by a more mature perspective.

Beneath his early vertically showy style, the essential message of Jesus was horizontal. His first teaching, after all, was the fact of beginning life "lying in a manger," and his whole work after that was a function of humility and simplicity, the last being made first. Simone Weil singles out two of his sayings as providing a critical key to all this. The first is, "When two or three are gathered in my name, there will I be also." She points out that Jesus said "two or three," and not fifty, a thousand, or millions, underscoring the fact that he intended his "way" to be a religion of intimates and friends, of "loving your neighbor," and not a mass movement. The second utterance is Jesus' reference to "thy Father who seest in secret." Again he stresses the individual heart and soul, eschewing the pomposity of public demonstrations of piety.

When Jesus was killed, he was crucified vertically, and that symbol became the primary icon of Christianity, along with the vertical "ascension" into "heaven," with only lip service given to the meaning of his arms being opened horizontally in an embrace of the world. Thus the churches founded in his name became a hideous parody of everything he lived and communicated, with the lilies of the field to be crushed by the weight of inquisitions, reformations, and the boots of Christian soldiers marching as to war.

The paradigmatic vertical political order is dictatorship, in which the top dog, or boss monkey, rules, and all social order is adjusted accordingly. In the supposed democracies, the boss is voted in, but the overall system is the same; also, the choice of candidate is controlled by those who, through wealth, exercise the real power in any case. This model is so ingrained that even when the "people," who are a horizontal mass, "rise up" to overthrow their oppressors, the first thing they do when they win is to construct a hierarchy of command.

This mechanism can be seen with clarity in the workings of the guru game, which is a voluntary dictatorship. What happens it that the guru appears and declares, or implies, that he (rarely she) has attained the "seventh level," and then calls others to join him, offering anyone who will listen the chance to "climb the ladder" also. The way to the top is standard: one must acknowledge the guru as "master" and live according to his laws. This process is known as "surrendering," and if one can successfully do that, the gates of love in the guru's heart open up and the follower is showered with blessings.

The principle is simple, but the practice takes a lifetime. Complete surrender is, statistically, very rare, which is not surprising since it involves total loss of ego. Most students, disciples and devotees spend years and decades struggling to make the grade, but extremely few get their enlightenment diploma or certificate of sanctity from the guru before he dies.

The result is that before long there are a lot of people milling about trying to slay their egos, and they have got to be organized into some kind of society so they can feed and shelter themselves. The form this takes is almost invariably a "community" with very strict rules for not only behavior but inner attitude. Traditional gurus use models that have been handed down, while radical gurus attempt to "create a culture." But no matter how they start out, these groups always end with the same social order—a hierarchy of concentric circles of power starting at the guru's feet and radiating downward.

In relation to the other dictatorships that flourish in the world, this game is practically harmless, although even here one often discerns the ripple of fascist muscle beneath the cloth of ritual compassion. But it is a far cry from the guru's claim to be representing the pinnacle of human development, and no guru has ever explained why, if they have "attained supreme enlightenment" or "come as agents of the divine," they can't create a social order more graceful than these relatively benign versions of the hard-edge secular dictatorships which rule the species. Probably because to do so would be to admit that the guru game, for all its values and virtues, is an inherently limited model. It's the best that the vertical understanding can produce by way of a social order, but it's just not good enough.

No one has spoken more forcefully and continuously against the guru game than J. Krishnamurti. For half a century he has traveled the globe doing a Shiva's dance of destruction, razing all the temples and teachings of the past five thousand years. But he has steadfastly refused to go beyond the parameters of his "negative intelligence." He is to the spiritual tradition what Einstein is to classical physics, the last of the old guard. He levelled all that went before but is incapable of envisioning what might come after, in the same way that Einstein balked at the implications of the new physics right up until his death.

Krishnamurti has a point in denouncing all the methods, systems and worldviews that spiritual teachers have produced, but he has not grasped their unifying principle, that of verticality, so he has missed the possiblity that a horizontal approach to these matters would allow the individual to wander the "pathless land" while still sharing a vision and technique with others. Also, he himself has an extremely vertical character structure, exemplified by his aversion to physical affection, his sneering at imagination and all forms of playfulness, and his forbidding any expression of mirth during his talks. So he has become stuck and, year after year he repeats the same injunctions to aging fans, chiding them for not having made any breakthrough, but categorically refusing to discuss how such a breakthrough might be made.

"The way we have lived over millions of years," he says, "has been the repetition of the same process of acquiring knowledge, and acting from that knowledge. That knowledge and action is limited. That limitation creates problems, and the brain has become accustomed to solving problems which knowledge has repeatedly created. The brain is caught in that pattern and we are saying that the pattern will never, in any circumstances, solve our human problems. Obviously, we have not solved them up to now. There must be a different, a totally different movement, which is a non-accumulative, perception-action."

So near and yet so far. This non-accumulative perception-action is precisely the essence of horizontal understanding. And what can the "totally different movement" be other than lying down? Krishnamurti rejects any effort to translate his notions into pragmatic terms, and he would die before letting his teaching be connected to any vision; but, in retrospect, his mission, ironically, may turn out to have been something like that of a Moses for the horizontal paradigm.

While this notion has its humorous aspect, there actually has been an emergence of a new breed of spiritual master and teacher appearing in the world since the Krishnamurtian revolution. Although their respective realizations and understandings are the same as their counterparts from the earlier epoch, they employ an entirely different semantic. With the onset of global communication and the cross-fertilization of cultures and traditions, their ancient message is cast in planetary terms which transcend parochial divisions. The foremost of these are in the United States and include the home-grown teacher Stephen Gaskin, the sufi siddha Lee Lozowick, and the aesthetic syncretist Bhagwan Shree Rajneesh.

The shift to horizontal primacy in spiritual practice involves more than attitude and form, however, and its structural foundation has been uniquely described by the radical transcendentalist Da Free John. In the Introduction to "The Heart is the Self," an essay on Sri Ramana Maharshi, the Editors of Laughing Man magazine give this synopsis of Da Free John's definition of "the heart."

"The traditional model of spiritual practice is the vertical path: the ascent to ever higher and subtler regions until the final, unsurpassable Goal is attained. In the Judeo-Christian tradition, the primary image of ascent is Jacob's ladder, which has its base on earth and its topmost rungs in heaven...

"However, when this picture is regarded in the light of the traditional esotericism of the East, the ladder is seen to be an archetypal description of the central nervous system of Man, or the 'Energy' which it reflects. In this view, the spinal column and brain form the frame of an esoteric ladder of Energy (the Current of Life) that is felt to rise into subtler and subtler realms of experience.

"This esoteric account of the nervous system and its underlying subtle 'energetic' functions is given its clearest expression in the *cakra* model of Hinduism and Tantric Buddhism. In this model, the spinal axis is seen to be constituted of a hierarchical series of (generally seven) energy centers ascending from the base of the spine, or the perineum, to the crown of the head...

"Traditionally, these *cakras*, or centers of consciousness, are connected by the three subtle 'pathways' or 'currents'—*ida, pingala,* and *sushumna*. *Ida* and *pingala* are a polarized pair of currents, or the two poles of a single current. They twine about the spinal axis meeting and crossing at the *cakras*... The third current, *sushumna*, represents the equanimity, harmony, or union of the two polar currents of *ida* and *pingala*. It is the central current that runs between the twin currents, coincident with the core of the spinal line, while the currents of *ida* and *pingala* are associated with the polarized play of opposites in the gross physical dimension. It is the vehicle of the more advanced or esoteric processes of traditional yogic and mystical spirituality, in which attention is directed inward and upward to the light contacted at the topmost center.

"However, in the ultimate extreme of ascent, the Life-Current does not terminate above or at the crown of the head, as most traditions teach. The traditional view is incomplete. As Master Da Free John explains, the Current falls in an S-shaped curve into the right side of the heart, the root of attention. Thus, there exists a third vertical structure beyond the twin currents of *ida* and *pingala*, and the single current of *sushumna*. This S-shaped current that runs between the center at (or above) the crown of the head and the right side of the heart is the vertical structure corresponding to the causal, formless dimension of thoughtless, bodiless pure awareness.

"As Master Da Free John points out, within this description of the vertical structure of the body-mind, the heart plays a central role. It is the dividing point between what is above and subtle and what is below and gross. The three centers below the heart are connected with the various functions of the gross physical being: vital-physical, emotional-sexual, and the verbal-intentional activity. The three centers above the heart are associated with the functions of the ascended, subtle being: mental (verbal), higher mental (comprehending), and supramental (intuitive-realizing). The heart center itself is the feeling, psychic core of the whole body, expressing the unity of the body-mind.

"However, the heart, the central focus of the two coils of the being, is not to be identified exclusively with the *anahata cakra*. The Teaching of Master Da Free John provides a unique decription of the 'heart,' as a complex *horizontal* structure of the body-mind. The 'heart,' which is not to be equated simply with the physical organ, is the 'horizontal foundation' of Man. As such, it is senior to the vertical structures of *ida, pingala, sushumna*, and the S-shaped current descending to the right side of the heart. Just as the vertical structures are evolutionary, dualistic, and linear—ascending toward subtler and subtler theatres of experience—the horizontal structure of the heart is prior, unitary, and central."

After a spell in Death Valley I was left with a powerful impression of the horizontality of the planet's surface. On a large scale, the only vertical things are mountains, and even they climb along the diagonal. Also, mountain ranges are horizontal across the horizon of their peaks.

With horizontality, there is emptiness. It is as though nothing happens. There is light and dark, life and death, and all the perceptions, feelings and behaviors of existence in awareness; but as palpably real as it is—the scorching sun mocking any fancy flights of abstraction—one feels as though in a dream: the dream of nature, where things mysteriously come into being and pass out of being, leaving no trace of their dance. The skip of a sidewinder across sand dunes, the shimmerings of heat waves above the salt flats, and the ancient alluvial flows are all simply different forms of vibration, without status or significance.

Only when the upright creature, the human, arrives, and erects a building or writes a poem, is there an "event." Some humans, like the Shoshoni, kept the event harmonious with the spirit of the place. Others, like the first white settlers, went mad and set immediately to work in the lethal heat, digging holes and building fires, making mines and smelters. The current inhabitants and visitors have struck a balance beween the Indians and the industrialists; human intrusion is evident but kept to a minimum so that the desert is almost entirely pristine.

With such impressions fresh in one's mind, driving out of the dry regions, toward the coast, toward Los Angeles, verticality reasserts itself with a Koyaanisqatsiesque acceleration. One sees the power lines, advertising signs, an increase in airplanes and helicopters overhead, the buildings growing taller and taller, everything becoming more crowded, more mechanical. And in light of the grandeur and insouciance of horizontal nature, the vertical appears even more harsh and ugly than usual.

Then, in one of those unsettling visions it can be dangerous to have at sixty miles an hour, I saw the civilization we have constructed not just as a vehicle for food, clothing and shelter, but as a panicky attempt to leave a mark on the emptiness, to fix a meaning in the flux, to carve our initials in creation. In that instant, I literally did not register the artifacts of my own species as anything like the familiar currency of culture, but as congealed existential self-importance, a piece of the void with a slogan on its t-shirt.

Yet, because what I was driving though was indeed the manufacture of my species and the manifestation of my culture, and because I was unquestionably a card-carrying member of both species and culture, I could not deny or denounce it. Obviously, if it were not for the industrial-financial base, which represents the vertical aesthetic in its grossest and most vulgar form, there would not be the asphalt highways nor the Ford Broncos to ride them onto the rough dirt roads that end where what's left of real wilderness begins. Nor would there by the oases of motel rooms and restaurants for people getting stir-crazy in their R.V.s.

The question then became: how much vertical is necessary and how much is a "denial of death"? Up to a point, the upright posture is simply an evolutionary experiment, a radical tool for survival. But in the same way that the sabre-toothed tiger's tusks turned in on themselves, our verticality out-reached itself and came to dominate our being-in-the-world. And since, as Straus described, the upright posture produced a primal resistance, restlessness and ambivalence to begin with, the vertical was ultimately transformed into what Ernest Becker calls our "*causa sui* project," which is our heroic gesture of identity, our bid for immortality, our cosmic tumescence.

When one stands solitary in the desert, or on a sailboat out of sight of land, it is crushingly clear that no amount of creating or building is going to alter the nature of existence or the human condition one little bit. Civilization simply cannot cover over the abyss it is trying to span. But when one is caught once more in traffic on the Freeway, it is equally clear that this culture-act we perform with and for one another, however beautiful or ugly, is all we have to hang onto in the universe.

I agree with Becker that we need illusions to survive, and there is no criticism of the species for making the planet a stage and the cosmos a theatre, or for putting on costume and mask and making speeches beneath the stars. But because our other survival tool, the upright posture, is vertical, we have made the mistake of creating vertical illusions and a vertical life theatre.

The work of translation in the metaphoric field of lying down is to render our illusions and our life theatre horizontal, so as to complement our vertical posture and ground it in ease, understanding and quiet strength. Just as the function of position in the physical field is to align our bodies with the horizontal, and the work of exploration in the intimate field is to make our way of knowing horizontal.

All of these merge in the fourth field, that of the horizon, where the mystical meets the absurd.

Some notes . . .

The blurb on the jacket of a popular paperback reads, "What if the one lying next to you is lying?" That "lying" means speaking falsely as well as getting horizontal gives lying down an ironic twist. One might meditate on the ways in which the lies of the vertical enter the realm of rest, relaxation and intimacy; and consider which lies arise *only* when we lie down.

Vertical styles of socializing include the meeting, formal dinner or any other gathering where the appearance and pose are more important than the substance. Horizontal socializing is more process-oriented, and involves a relaxed attitude in which all aspects of the get-together are enjoyed, and not just the "business." Verticality focuses on a point and squeezes all available energy in a relentless drive to reach it, an excellent approach in emergencies, but why live life in a state of chronic emergency vibration?

To the degree that there is a hierarchy, we get what C.S. Lewis called the "lowerarchy," that is to say, the more we attempt to go beyond our condition, the more deeply are we mired in it. The solution is to rest on and in the condition easily, comfortably, to understand our existence from a horizontal viewpoint, not in terms of "levels."

When we are standing erect, only the soles of the feet are horizontal. That is why the zen master advised his student to "walk on."

The Russians will see Lying Down as the final proof of the West's decadence and exhaustion. Paranoid Westerners will see it as a Communist plot to keep the United States from "being Number One."

Lying down after physical exertion is very sweet; lying down after catharsis is pure rapture.

Lying down under the sheet, in the dark, alone, the door locked.. in that secret space then ask, "Who am I?" and "Is there God?"

If walking, standing, sitting and lying down are the "four dignities," then the "undignities" would include running, jumping, skipping, rolling, falling, diving, crawling, sliding, slipping, staggering, swooning, swimming, dancing, sexing, kneeling, crouching, slouching, being on all fours, miming, riding, playing games, and freaking out.

What if Luther, instead of his vertical challenge of "Here I stand," had said, "Here I lie, I can't do otherwise"? And then slipped his theses under the cathedral door instead of nailing them there? Might we not now have a wry and relaxed Protestantism instead of the stilted and often violent vehicle it has become?

Normality as fixed limits; neurosis as false limits; psychosis as virtual limits; maturity as organic limits; enlightenment as pure limit.

Creation of a morphogenetic field: when you lie down, lay the world down with you.

Bag the vertical. Lying down is awesome. Fer Shur-r-r-r-r.

Latitudes

"When thou liest down, thou shalt not be afraid; yea, thou shalt lie down and thy sleep shall be sweet." —*Proverbs 3:24*

"Come unto me all you that labor and are heavy laden, and I will give you rest." —Jesus

"O, what is more sweet than when the mind, set free from care, sets down its burden; and when spent with distant travel, we come back to our home and rest our bodies on the wished-for bed?" —Catullus

"My love is my weight." —Augustine

"Go placidly amid the noise and haste, and remember what peace there may be in silence." —*Desiderata*

"I have often said that all the unhappiness of humanity comes from not knowing how to remain quiet in a chamber."—Pascal

"The happiest part of a man's life is what he passes lying awake in bed in the morning. I have, all my life long, been lying til noon; yet I tell all young men, and tell them with great sincerity, that nobody who does not rise early will ever do any good." —Samuel Johnson

"The bed has become a place of luxury to me. I would not exchange it for all the thrones of the world." —Napoleon I

"Sleep faster, we need the pillows." —Yiddish saying

"I want a sofa, as I want a friend, upon which I can repose familiarly. If you can't have intimate terms and freedom with one and the other, they are of no good." —Thackeray

"It was not Death, for I stood up,
And all the Dead, lie down." —Emily Dickinson

"The bed, my friend, is our whole life. It is there that we are born, it is there that we love, it is there that we die." —Guy de Maupassant

"Live and lie reclined
On the hills like gods together, careless of
mankind." —Tennyson

"I loafe and invite my soul,
I lean and loafe at my ease observing a spear of summer grass."
—Walt Whitman

"Now that my ladder's gone,
I must lie down where all the ladders start,
In the foul rag-and-bone shop of the heart." —Yeats

"All of us are in the gutter, but some of us are looking at the stars."
—Oscar Wilde

"And we, who have always thought of joy
as *rising*, would feel the emotion
that almost amazes us
When a happy thing *falls*." —Rilke

"...and if you don't mind I think
I'll just lie this one out in my own way." —Kenneth Patchen

"We do not need the perfectly designed chair, we want a capacity for relaxation even on a bed of nails." —Quentin Crisp

"The key to enlightenment and serenity during the period of ten thousand visions is simply this: relax." —*The Psychedelic Experience*

"The lion and the lamb might lie down together, but the lamb won't get much sleep." —Woody Allen

"The slang term 'hip' comes from the opium trade. When you smoke you lie on your hip. When you say 'I'm hip,' that means you know the life." —Robert Mitchum

"There is a limit to human perception, and beyond this horizon the fruit of observation will be gleaned from other beings, research machines, or whatever. Progress may continue but we will somehow be staying behind." —Stanislaw Lem

"There's nothing left to talk about
Unless it's horizontally." —Olivia Newton-John

4 Horizon
The Field of the Mystic/Absurd

"There comes a critical moment when everything is reversed, after which the point becomes to understand more and more that there is something which cannot be understood."

—Kierkegaard

The fourth field is the limit to all fields—the horizon.

We have always lived with the horizon. When we peered from the edges of the forest onto the dangerous yet inviting plain, we must have been intrigued by the mystery of what lay beyond the far horizon. Impelled by hunger, curiosity, or perhaps some divine mischief, we descended the vertical trees to begin a horizontal journey as vertical creatures.

Once in the open, we had to continually scan the space around us, searching for food and water, looking out for predators. This required developing a way of seeing that is at once flexible and precise. The perfection of binocular vision took care of part of that need, but there also had to be a method of perceiving which produced the best chances of survival, and this was probably the technique of keeping the eyes resting lightly on the horizon, the way don Juan advises Castaneda to do during one of their exercises. This allows not only maximum peripheral vision but also establishes a reference for gestalting the area, so that any sudden movements or unusual shadows stand out immediately.

Then, there must have come a day when we became aware of a strange continuity: that as we moved about, although the topography and texture of the land changed beneath our feet, the horizon we gazed at and used as a marker never approached or receded. It was constant. That realization may have been our first conceptual discrimination, our first abstraction, our first *idea*.

This undoubtedly coincided with an advanced stage in the growth and complexification of the brain, when its capacity for pattern recognition had been developed. So, although that early biocomputer was given only one bit of information, it already had the circuits to begin reasoning from it. The concept of horizon may have been the single seed-crystal of consciousness that lit up the entire neural board, an event that probably had the feel of the moment in *2001* when the ape grasps the meaning of weapon.

Our next insight into the nature of the horizon came very much later, probably just at about the time of the transition from the "magical" period into farming. This was the discovery that the horizon is an illusion. We probably had not yet figured out that the Earth is a globe and the horizon a non-existent line produced by the planet's curving away beyond the range of sight, but because we'd learned how to create and use symbols, it dawned on us that the horizon "line" was as artificial as the ones we drew on the cave walls.

I imagine it was unsettling to realize that the very thing that had crystalized consciousness was an illusion, and it may explain why, with farming, there came a spate of boundary making. Almost as though we were nervous at having lost the very limit of perception that had defined the mind, we started creating limits. We built walls, political borders, language zones, cultural taboos, property lines, levels of authority, and finally, the ultimate distinction, the sense of "I," the ego, which delineates itself not only from nature but from other human beings and indeed from everything else in the universe. In short, we invented a world of imaginary limits and then imprisoned ourselves within them.

Whether we do it to ourselves, however, or whether nature does it to us, being confined by unreal boundaries is what fundamentally defines the human condition. And even being sophisticated enough to understand this does not change it. In fact, the more we come to know about the universe, the more ironic does the situation become. We have now seen photographs of the Earth taken from space and can smile indulgently at our ancestors who thought themselves limited by the horizon. But when we use our superior technology and advanced knowledge to peer about us today, everywhere we look we find ourselves facing...another horizon.

In astronomy, for example, we discover implacable limits. First is the "space horizon," describing those galactic regions receding so rapidly from us that their light won't reach us until long after our sun has exploded and vaporized the solar system. Then, of course, the black holes and their "event horizon," from whose bourne no traveler returns. And now it seems we must be on the lookout for "domain walls," invisible sheets of energy that divide the cosmos into different realms, and which will shred us if one gets too close.

Going from large to small, we enter the realm of quantum slipperiness where all we can know of reality is our statistical description of it, and Alice in Wonderland visualizations of the way things work at the most elementary level. Electrons thumb their noses at us and particles are discovered whose existence cannot be assumed either before or after they are perceived. The invisible world consitutes a horizon that continually mocks all efforts to penetrate it, or even to prove that it is there at all.

Our two most popular cosmologies reinforce the sense of limit. The first is the notion of an infinitely old universe, self-existing, without beginning or end, within which our local universe is a miniscule flash in the pan. In this scenario, our limitation is very limitlessness: we can never know more than an infinitesimal bit of the whole thing. No matter how far we see, we will always be peering myopically into unimaginable vastness.

The second cosmological vision derives from the notion of spontaneous emergence, or big bang. Here, time-space and energy-matter appear from nothing, or from something about which we can know nothing. The horizon is a primordial nanosecond erupting as a bizarre singularity, without any context we can comprehend. In this scenario, we can learn a great deal about what happened since then, but can never know what it all came out of. Also, this kind of universe ends badly, in one of two ways: either it drifts apart into eternal darkness and cold, or collapses in upon itself in an implosion of absolute heat—no mean horizon in either case.

All of this suggests a paraphrase of Gödel's theorem to the effect that any perception pushed to its limit always ends in a horizon. This is an idea that has been understood by at least some thinkers over the past few thousand years, and forms the basis for all absurdist outlooks, as well as the springboard for mystical claims. The absurdist greets limitation with irony while the mystic takes it as a challenge. In the esoteric disciplines of the major religions, it was accepted that the "eye of flesh" and the "eye of reason" were limited, but a "third eye," was invoked, the "eye of contemplation," by which one could transcend all horizons.

But upon examination, this claim is found to be misleading: insofar as it is a claim. If a mystic remains silent, there can of course be no argument, no discussion. But when such a person articulates, or "draws," his or her mystic vision, the very line of expression and communication constitutes its own horizon. Also, if one sifts through all the sayings of the mystics in and around the great traditions, their depictions of the Ultimate can be classified into two, mutually exclusive, fables—that of the Creator and that of the Emptiness.

In the first story, some being—God—decides to make the world for reasons we don't understand and may not even approve of. We may attempt to communicate with this being, or imagine that it is directing our lives or even playing favorites by sending messengers to certain select groups of people, but in all cases it is something intrinsically Other. The horizon is the line where we leave off and it continues.

The second tale tells of primal unqualified consciousness, a kind of brimming emptiness in which everything appears and disappears mysteriously for no reason whatsoever, and to absolutely no effect. This isn't very different from the contemporary physicist's postulation that space-time arose from random fluctuations in a primordial nothingness. The horizon here is the irreducible mystery of why there is anything rather than nothing at all.

So, while we've become far more sophisticated than our ancestors on the early plains, our condition is the same—we find ourselves gazing at horizons. Like them, we furrow our brows and scratch our heads and wonder what might lie beyond. We build cyclotrons and construct systems of metaphysics, we meditate and pray, all with the hope of going beyond whichever horizon is obsessing us.

Why do we continue to strive so hard? Do we believe that there is an alternate reality waiting to open its doors to us? Are we hoping that we will at last come upon a territory from which some kind of inconceivable illumination or release or salvation will appear? If, in our probes of the universe and of our minds, we keep finding more and more horizons, why not lay the whole game of horizon-hopping to rest?

The answer to that question has been the substance of this book: that our urge to "penetrate" the mystery is simply the reflex habit of the momentum of the vertical thrust that has defined us as a species. And like all habit, it has made us stupid, preventing us from seeing the obvious truth that we can't go beyond the horizon precisely *because* it *is* an illusion. Behind the visual horizon there is only more planet, behind death there is only more life, and behind anything we have come up against in the universe there has only been more universe.

The message we might learn from all this is summed up in a line from an ancient scripture to the effect that "this is the other shore." But we are kept from that refreshingly arresting conclusion not only by the force of vertical acceleration, but by pride. This is traditionally considered the original sin, born with Lucifer's words, "I will not serve," a phrase that in human terms might read, "I will not accept limitation." The Angel of Light would not submit to God in human form and for his rebellion was banished to hell where he would have to spend all his time with humans. We will not submit to a boundary that is an illusion, and for our rebellion are sentenced to self-consciousness, which has the most unreal and painful borders of all.

The way out of this cage, I have been suggesting, is by lying down in its four fields. In the first three fields this is done through physical posture, through awakening awareness and intimacy as mediums of knowing, and through translating our life theatre and illusions into a horizontal worldview. In the fourth field, the vehicle is a direct and complete acceptance of the horizon as absolute. This means accepting that no matter how far we extend our perception, in whatever dimension, at whatever level, it will always be defined by a horizon, a "line" that does not exist and yet which keeps us in our place.

With this philosophic understanding there must also be a moral transformation: the pride of the vertical must be dissolved in the humility born of making peace with the meaning of the horizon. This is not the Sunday school humility of pious attitudes but a clear, sane, adult comprehension that we are not superior to anything on this planet or in this universe, and that the very notion of superiority is our basic sickness as a species. All of this is decidedly difficult for a creature that likes to bill itself as the "crown of creation," but our survival and evolution depend on it. As Eliot put it, "The only wisdom we can hope to acquire/Is the wisdom of humility: humility is endless."

And with this comes the final paradox. For, if the horizon is everywhere and defines everything, when we lie down and enter the horizon, then we are everywhere and define everything. Not by climbing higher or probing deeper, but by letting ourselves be absorbed into that which is our very ground. It is in the ironic acceptance of the absurdity of having reality shaped by illusion that we mystically transcend the entire show.

Then there is nothing other than living each day, doing what work we must, loving the best we can, and looking upon all that is with a sense of wonder.

Which makes the new paradigm a very old story after all.

Confession

I was first introduced to lying down as a practice twenty-three years ago when I was an undergraduate and one of my psychology professors invited me to participate in "an experiment in relaxation" she was conducting with some of her students. Once a week, eight or nine of them had been gathering to lie on the rug in her studio for a few hours while she instructed them in ways to contact their tensions and "breathe through" them.

The first time I went, after no more than five minutes of the process, I was overwhelmed by a wave of sadness, and began sobbing uncontrollably, a fit of crying that went on for some fifteen or twenty minutes. When it subsided, I felt a sudden and profound release, followed by a virtual overload of memory and insight. It was a naive and raw satori.

I studied with her for five years, and with her off-campus colleague, a Stanislavskian acting coach who used lying down as his principal technique. During that time we uncovered an extraordinary wealth of material on the psychophysical and aesthetic dimensions of the practice, but we had no satisfactory context within which to understand what was happening. It was a dark time intellectually, just after the Eisenhower years, with the smoke still curling over the dumpsters where Reich's books had been burned.

The context was right around the corner, however. In short order there appeared zen, LSD, Esalen, national radical politics, the meta-sexual revolution, and all the other eruptions of our recent renaissance. I got caught up in all that, and the realm of lying down I'd been exploring faded into the background. Also, my two teachers dissolved their partnership and moved into different areas, the acting coach into primal scream and the psychologist into faith healing.

But I came cross the practice many times since, encountering it in many other schools and situations through the turbulent Sixties and the miasmic Seventies, both formally and informally, and as a teacher as well as a student. After a while I came to see it as a leitmotiv to my studies, travels and writing. It became a kind of personal esoteric discipline.

Then, last year, around the time of my forty-fifth birthday, as I was undergoing a heavy transformation in my life, I began to find in lying down an urgency, a quickening, that the posture had never evinced before. One night, as I was talking about this with two friends, in the midst of the conversation I seemed to turn a corner in my mind. In that instant I realized that lying down wasn't the subplot of the text of my life story, but the guiding principle and source of illumination of all my work for almost a quarter of a century. I saw that it had been the real theme, informing everything else all along.

One of the friends in that conversation suggested that I do a book on the subject, something that had never occurred to me before. She took the idea to "a friend who is a publisher," and to my surprise a contract and advance were forthcoming almost at once. Thus, before I'd fully considered the implications of what I'd committed myself to, I found myself involved in this odd business of using a vertical posture and method to communicate a message about the horizontal. For, as I noted in the Introduction, I am still sitting at the typewriter as I compose these sentences about lying down.

The entire process has been a fascinating exercise in both postural awareness and conceptual gymnastics. And even more important, the work of bringing into figure what had been ground brought about an alchemical transformation in my relationship to lying down. The book was a crucible in which the vertical was pulverized, and mixed with the horizontal essence, to produce a different quality of being.

There were times in this literary laboratory when I felt that I was betraying myself by translating something secret and sacred into a form suitable for public consumption. Communicating experience and idea gives them one kind of power, but at the expense of a certain fragility and beauty.

At other times I was seized by messianic fervor. I imagined that upon publication of the book all the Buddhists in the world would abandon sitting en masse and convert to lying down as their primary vehicle, with a domino effect on the rest of humanity, producing global peace within a decade. Another scenario had a hundred people gathering to form a group mind, along the lines suggested by David Bohm. Using lying down in its four fields as the medium of communication and communion, they become a force for compassion and intelligence as powerful as the terminally vertical Nazi group mind was for stupidity and violence. At another time I was convinced that there would be a sudden, spontaneous emergence of a lying down "movement" which sweeps the country, but in a horizontal manner, as a complex of networks, and not vertically as an organization or political party.

And at moments of startling lucidity in the writing of this vision, it was clear to me that there is, and always has been, a split in the species between vertical peoples and horizontal peoples, and that the difference between them is not merely cultural but mutational. From the very first, as the species went upright, it divided into two strains: both stood erect, but one sold its soul to the vertical while the other never forgot its spiritual connection to the horizontal life base on the planet. The second group has tended to live peacefully and away from the highways of the aggressive hyper-verticals, but has inevitably been attacked and conquered by them anyway, as has happened with all native peoples, the olde religion, and the true underground. Ironically, now that the movement to the horizontal has become an evolutionary imperative even for the mechanistic verticals, these "throwbacks" constitute the real leadership of humanity.

Thus, in the nine months of sitting at the typewriter to tell the story of lying down, I went from fear that I was vulgarizing my most precious treasure, through manic euphoria over images of a horizontally transformed humanity, and into a sober *real politik* dualism, and all of it connected to an alchemical process the result of which I could not predict. Added to that, at this point, is the author's perennial concern as to how the book will be received.

All in all, it has been quite an adventure. I hope that, in whatever posture you read the work, and whatever you make of its claims to articulate the new paradigm, it has been an enjoyable and even healing voyage. If nothing else, I trust that it has awakened some vital sense of postural awareness and its dignities, and that is has given you a more dimensional appreciation of lying down—something that we will all be doing every day and every night for the rest of our lives, and at our death.

The fields of lying down are always open to anyone who wants to enter. They are at our feet, in our hearts, and in the world around us, as far as the horizon in any direction, and into the horizon, and so beyond.

About the Author

My maternal grandparents, Luigi and Angelica Freda, and my paternal grandparents, Maria and Ferdinando Vasquez-d'Acugno, all emigrated from Naples in the 1880s to settle in the United States. My parents, Guglielmo and Angelina, were married during the Great Depression, and I was born on November 6, 1937 in New York City.

Ferdinand William Vasquez (1937-1951)

I grew up an only child in the Italian immigrant neighborhoood of East Harlem in Manhattan and attended our Lady of Mount Carmel grammar school, run by nuns, one of whom was my first formal mentor. My major influences were the medieval order of the Church and Mafia, the tribal lessons of the streets, and discovering the magic of reading.

When I was thirteen my parents, in the spirit of cultural assimilation, had my name legally changed.

Fred Vassi (1951-1970)

Leaving the neighborhood for the first time, I entered Regis, the Jesuit high school, but left after two years and graduated from Saint Ann's Academy. I matriculated at Iona College, and in the middle of my sophomore year dropped out to enlist in the Air Force, where I was picked for Intelligence work. I was sent to the Institute of Far Eastern Languages at Yale University for a year-long intensive course in Mandarin Chinese, with the next three years spent in Korea and Japan working as a translator. While in Japan I married, though was to be divorced two years later, and was introduced to zen at the Rock Garden of Ryoanji in Kyoto.

After my discharge from the service I resumed academic studies, getting a B.A. from Brooklyn College and beginning a master's program in Psychology at the New School for Social Research. Supporting myself by teaching fifth grade in the public school system, I completed fifteen credits before dropping the course.

I next took a position as Editor of the house organ at Encyclopedia Americana, and a year later joined the staff at the Institute for Holistic Personality Research where we designed psychophysical experiments in body awareness based on the work of Else Gindler and Wilhelm Reich. Also, during this period, I studied Political Economy at the Center for Marxist Studies, esoteric disciplines at the Gurdjieff Foundation, therapy at the Institute for Bioenergetic Analysis and the Stanislavski Method at the Theatre of Encounter.

In mid-1967 I went to California to teach a class in whole-body awareness at San Francisco State Experimental College, and a workshop in body imagery at the R.D. Laing Center at Agnew State Hospital in San Jose. During this time I participated in Stephen Gaskin's Monday Night Class, began a seven-year study of the work of J. Krishnamurti, and experimented with psychedelics.

In late 1969 I returned to New York City and co-founded Raindance Corporation, a metatheatrical video company, and Radical Software, a media interface quarterly.

Marco Vassi (1970-   )

Using a new first name, I began a career as a professional writer. Within thirteen years I had a dozen books and several hundred stories and essays published, with novels translated into Japanese, German and Italian.

During this time I also worked as Editor of Penthouse Press and Executive Editor of Variations magazine for several years, and gave talks and workshops at Anthos, a growth center.

From 1976 to 1981 I conducted an investigation of contemporary erotic possibilities, embodying and articulating the full range of metasexual forms, including heterosexual neo-monogamy, gay romance and promiscuity, male lesbianism, bisexual double-coupling, the triad, the swing, the orgy, and the sadomasochistic ritual.

Concurrently, I studied with a number of radical and traditional spiritual masters and teachers, most notably Bawa Muhaiyaddeen, Lee Lozowick, Eido Shimano, Stanley Fisher, and Hilda Charlton, and tested the communities of Bhagwan Shree Rajneesh and Da Free John.

In late 1983, with the completion of *Lying Down,* this cycle of my life came to a close and I began a new phase of work.